STAAR Subject Test

Mathematics Grade 4

Student Practice Workbook

+ Two Full-Length STAAR Math Tests

Math Notion

www.MathNotion.com

STAAR Subject Test Mathematics Grade 4

STAAR Subject Test Mathematics Grade 4

STAAR Subject Test Mathematics Grade 4

Published in the United State of America By

The Math Notion

Web: WWW.MathNotion.com

Email: info@Mathnotion.com

Copyright © 2021 by the Math Notion. All rights reserved. No part of this publication may be reproduced, stored in a retrieval system, or transmitted in any form or by any means, electronic, mechanical, photocopying, recording, scanning, or otherwise, except as permitted under Section 107 or 108 of the 1976 United States Copyright Ac, without permission of the author.

All inquiries should be addressed to the Math Notion.

ISBN: 978-1-63620-097-2

> STAAR Subject Test Mathematics Grade 4

The Math Notion

Michael Smith has been a math instructor for over a decade now. He launched the Math Notion. Since 2006, we have devoted our time to both teaching and developing exceptional math learning materials. As a test prep company, we have worked with thousands of students. We have used the feedback of our students to develop a unique study program that can be used by students to drastically improve their math scores fast and effectively. We have more than a thousand Math learning books including:

- SAT Math Prep
- ACT Math Prep
- SSAT/ISEE Math Prep
- Mathematics Prep Grade 3 to 8
- Common Core Math Prep
- many Math Education Workbooks, Study Guides, Practice and Exercise Books

As an experienced Math test preparation company, we have helped many students raise their standardized test scores—and attend the colleges of their dreams: We tutor online and in person, we teach students in large groups, and we provide training materials and textbooks through our website and through Amazon.

You can contact us via email at:

info@Mathnotion.com

STAAR Subject Test Mathematics Grade 4

Get the Targeted Practice You Need to Ace the Texas STAAR Math Test!

STAAR Subject Test Mathematics Grade 4 includes easy-to-follow instructions, helpful examples, and plenty of math practice problems to assist students to master each concept, brush up their problem-solving skills, and create confidence.

The STAAR math practice book provides numerous opportunities to evaluate basic skills along with abundant remediation and intervention activities. It is a skill that permits you to quickly master intricate information and produce better leads in less time.

Students can boost their test-taking skills by taking the book's two practice STAAR Math exams. All test questions answered and explained in detail.

Important Features of the 4th grade STAAR Math Book:

- A **complete review** of STAAR math test topics,
- Over 2,500 practice problems covering all topics tested,
- The most important concepts you need to know,
- Clear and concise, easy-to-follow sections,
- Well designed for enhanced learning and interest,
- Hands-on experience with all question types,
- **2 full-length practice tests** with detailed answer explanations,
- Cost-Effective Pricing,

Powerful math exercises to help you avoid traps and pacing yourself to beat the Texas STAAR test. Students will gain valuable experience and raise their confidence by taking 4th grade math practice tests, learning about test structure, and gaining a deeper understanding of what is tested on the STAAR math grade 4. If ever there was a book to respond to the pressure to increase students' test scores, this is it.

WWW.MathNotion.COM

… So Much More Online!

- ✓ FREE Math Lessons

- ✓ More Math Learning Books!

- ✓ Mathematics Worksheets

- ✓ Online Math Tutors

For a PDF Version of This Book

Please Visit WWW.MathNotion.com

STAAR Subject Test Mathematics Grade 4

Contents

Chapter 1 : Place Values and Number Sense **11**
   Place Values 12
   Comparing and Ordering Numbers 13
   Numbers in Standard Form 14
   Numbers in Word Form 15
   Roman Numerals 16
   Rounding Numbers 17
   Odd or Even 18
   Answers of Worksheets 19

Chapter 2 : Whole Number Operations **21**
   Adding Whole Numbers 22
   Subtracting Whole Numbers 23
   Multiplying Whole Numbers 24
   Dividing Hundreds 25
   Long Division by Two Digits 26
   Division with Remainders 26
   Rounding Whole Numbers 27
   Whole Number Estimation 28
   Answers of Worksheets 29

Chapter 3 : Fractions **31**
   Simplifying Fractions 32
   Like Denominators 33
   Compare Fractions with Like Denominators 35
   More Than Two Fractions with Like Denominators 36
   Unlike Denominators 37
   Ordering Fractions 39
   Denominators of 10, 100, and 1000 40
   Multiplying Fractions 42
   Answers of Worksheets 43

Chapter 4 : Mixed Numbers **46**
   Fractions to Mixed Numbers 47
   Mixed Numbers to Fractions 48
   Add and Subtract Mixed Numbers 49
   Multiplying Mixed Number 50
   Answers of Worksheets 51

STAAR Subject Test Mathematics Grade 4

Chapter 5 : Decimals **53**
    Graph Decimals 54
    Adding and Subtracting Decimals 55
    Multiplying and Dividing Decimals 56
    Rounding Decimals..57
    Comparing Decimals...58
    Answers of Worksheets 59

Chapter 6 : Patterns and Algebraic Thinking **61**
    Repeating Patterns 62
    Growing Patterns 63
    Patterns: Numbers..64
    Finding Rules 65
    Algebraic Word Problems 66
    Evaluating Outputs 67
    Answers of Worksheets 68

Chapter 7 : Measurement **71**
    Reference Measurement Units 72
    Metric Length Units 73
    Customary Length Units 73
    Metric Capacity Units 74
    Customary Capacity Units 74
    Metric Weight and Mass Units 75
    Customary Weight and Mass Units 75
    Time 76
    Money Amounts..77
    Money: Word Problems 78
    Answers of Worksheets 79

Chapter 8 : Geometric **81**
    Identifying Angles 82
    Estimate Angle Measurements..83
    Measure Angles with a Protractor 84
    Polygon Names 85
    Classify Triangles 86
    Parallel Sides in Quadrilaterals...87
    Identify Rectangles 88
    Perimeter: Find the Missing Side Lengths 89
    Perimeter and Area of Squares...90
    Perimeter and Area of rectangles 91

WWW.MathNotion.Com

STAAR Subject Test Mathematics Grade 4

    Find the Area or Missing Side Length of a Rectangle...92
    Area and Perimeter: Word Problems..93
    Volume of Cubes and Rectangle Prisms 94
    Answers of Worksheets 95

Chapter 9 : Data Graphs, and Statistics...97
    Bar Graph 98
    Tally and Pictographs 99
    Dot plots..100
    Line Graphs...101
    Stem-And-Leaf Plot 102
    Coordinate Plane 103
    Answers of Worksheets 104

Chapter 10 : Three-Dimensional Figures 106
    Identify Three-Dimensional Figures 107
    Count Vertices, Edges, and Faces 108
    Identify Faces of Three-Dimensional Figures 109
    Answers of Worksheets 110

Chapter 11 : Symmetry and Transformations..111
    Line Segments 112
    Identify Lines of Symmetry 113
    Count Lines of Symmetry 114
    Parallel, Perpendicular and Intersecting Lines 115
    Answers of Worksheets 116

Chapter 12 : STAAR Math Practice Tests 117
    STAAR GRADE 4 MAHEMATICS REFRENCE MATERIALS 119
    STAAR Practice Test 1 121
    STAAR Practice Test 2 133

Chapter 13 : Answers and Explanations...145
    Answer Key...145
    Practice Test 1 147
    Practice Test 2 151

WWW.MathNotion.Com

STAAR Subject Test Mathematics Grade 4

Chapter 1 : Place Values and Number Sense

Topics that you'll learn in this chapter:

- ✓ Place Values,
- ✓ Compare and Ordering Numbers,
- ✓ Number in Standard Form
- ✓ Numbers in Word Form,
- ✓ Roman Numerals,
- ✓ Rounding Numbers,
- ✓ Odd or Even,

STAAR Subject Test Mathematics Grade 4

Place Values

✎ Write numbers in expanded form.

1) Sixty–two ___ + ___

2) fifty–six ___ + ___

3) thirty–one ___ + ___

4) forty–five ___ + ___

5) twenty–eight ___ + ___

✎ Circle the correct choice.

6) The 6 in 56 is in the

   Ones place   tens place   hundreds place

7) The 2 in 25 is in the

   Ones place   tens place   hundreds place

8) The 9 in 918 is in the

   Ones place   tens place   hundreds place

9) The 3 in 537 is in the

   Ones place   tens place   hundreds place

10) The 9 in 289 is in the

   Ones place   tens place   hundreds place

STAAR Subject Test Mathematics Grade 4

Comparing and Ordering Numbers

✎ Use less than, equal to or greater than.

1) 31 _____ 33   8) 42 _____ 36

2) 57 _____ 49   9) 55 _____ 55

3) 92 _____ 88   10) 57 _____ 75

4) 76 _____ 67   11) 28 _____ 38

5) 43 _____ 43   12) 19 _____ 15

6) 54 _____ 46   13) 82 _____ 90

7) 97 _____ 88   14) 78 _____ 84

✎ Order each set numbers from least to greatest.

15) – 18, – 22, 28, – 17, 4   ___, ___, ___, ___, ___, ___

16) 19, –36, 11, – 12, 5   ___, ___, ___, ___, ___, ___

17) 27, – 56, 20, 1, – 27   ___, ___, ___, ___, ___, ___

18) 26, – 96, 2, – 26, 87, –75   ___, ___, ___, ___, ___, ___

19) –10, –71, 70, –26, –59, –39   ___, ___, ___, ___, ___, ___

20) 88, 4, 38, 7, 78, 9   ___, ___, ___, ___, ___, ___

21) 84, 14, 24, 0, 35, 22   ___, ___, ___, ___, ___, ___

Numbers in Standard Form

✎ Write the number in standard form.

1) 14 million 154 thousand 8
   _____

2) 89 million 15 thousand 798
   _____

3) 97 million 5 thousand 8
   _____

4) 124 million 2 thousand 2
   _____

5) 50 billion 3 million 5 thousand 4
   _____

6) 34 billion 45 million 578
   _____

7) 94 billion 21 million 51 thousand
   _____

8) 58 billion 708 thousand 120
   _____

9) 59 billion 54 thousand 86
   _____

10) 74 billion 354 thousand 158
    _____

11) 7 billion 13 million 12 thousand 7
    _____

12) 72 billion 450 million 658
    _____

13) 398 million 67 thousand 128
    _____

14) 24 billion 54 million 9 thousand 32
    _____

15) 795 million 458
    _____

16) 38 billion 2 million 54 thousand 9
    _____

STAAR Subject Test Mathematics Grade 4

Numbers in Word Form

✎ Write each number in words.

1) 372 _____

2) 605 _____

3) 550 _____

4) 351 _____

5) 793 _____

6) 647 _____

7) 3,219 _____

8) 5,326 _____

9) 2,842 _____

10) 4,691 _____

11) 5,531 _____

12) 7,360 _____

13) 2,532 _____

14) 8,014 _____

15) 11,242 _____

Roman Numerals

✎ Write in Romans numerals.

1	I	11	XI	21	XXI
2	II	12	XII	22	XXII
3	III	13	XIII	23	XXIII
4	IV	14	XIV	24	XXIV
5	V	15	XV	25	XXV
6	VI	16	XVI	26	XXVI
7	VII	17	XVII	27	XXVII
8	VIII	18	XVIII	28	XXVIII
9	IX	19	XIX	29	XXIX
10	X	20	XX	30	XXX

1) 11 _____ 2) 21 _____

3) 24 _____ 4) 16 _____

5) 27 _____ 6) 29 _____

7) 12 _____ 8) 28 _____

9) 15 _____ 10) 20 _____

11) Add 16 + 14 and write in Roman numerals. _____

12) Subtract 34 – 5 and write in Roman numerals. _____

STAAR Subject Test Mathematics Grade 4

Rounding Numbers

Round each number to the underlined place value.

1) 3,<u>7</u>93

2) 3,<u>8</u>76

3) 3,4<u>5</u>2

4) 7,1<u>9</u>3

5) 5,2<u>7</u>8

6) 1,4<u>7</u>7

7) 8,<u>3</u>13

8) 24.<u>6</u>8

9) 8<u>4</u>.92

10) 71.<u>3</u>4

11) 66<u>4</u>.7

12) <u>9</u>,135

13) 15.3<u>8</u>1

14) 4,<u>5</u>21

15) 3<u>6</u>.50

16) 4,<u>8</u>19

17) 6,6<u>8</u>5

18) 2,5<u>3</u>8

19) 73.<u>6</u>2

20) 16,<u>5</u>27

21) 2<u>9</u>.720

22) 12,3<u>6</u>6

23) 31,<u>7</u>29

24) 7,8<u>3</u>8

STAAR Subject Test Mathematics Grade 4

Odd or Even

✎ Identify whether each number is even or odd.

1) 18 _____   7) 80 _____

2) 27 _____   8) 53 _____

3) 21 _____   9) 58 _____

4) 17 _____   10) 98 _____

5) 67 _____   11) 49 _____

6) 76 _____   12) 113 _____

✎ Circle the even number in each group.

13) 52, 11, 35, 73, 5, 29   15) 33, 45, 86, 59, 63, 87

14) 13, 15, 113, 87, 71, 18   16) 55, 32, 79, 51, 21, 83

✎ Circle the odd number in each group.

17) 54, 36, 48, 76, 71, 100   19) 58, 92, 25, 78, 76, 50

18) 32, 56, 40, 74, 98, 67   20) 89, 12, 88, 42, 48, 120

STAAR Subject Test Mathematics Grade 4

Answers of Worksheets

Place Values

1) 60 + 2
2) 50 + 6
3) 30 + 1
4) 40 + 5
5) 20 + 8
6) ones place
7) tens place
8) hundreds place
9) tens place
10) ones place

Comparing and Ordering Numbers

1) 31 less than 33
2) 57 greater than 49
3) 92 greater than 88
4) 76 greater than 67
5) 43 equals to 43
6) 54 greater than 46
7) 97 greater than 88
8) 42 greater than 36
9) 55 equals to 55
10) 57 less than 75
11) 28 less than 38
12) 19 greater than 15
13) 82 less than 90
14) 78 less than 84
15) –22, –18, –17, 4, 28
16) –36, –12, 5, 11, 19
17) –56, –27, 1, 20, 27
18) –96, –75, –26, 2, 26, 87
19) –71, –59, –39, –26, –10, 70
20) 4, 7, 9, 38, 78, 88
21) 0, 14, 22, 24, 35, 84

Numbers in Standard Form

1) 14,154,008
2) 89,015,798
3) 97,005,008
4) 124,002,002
5) 50,003,005,004
6) 34,054,000,578
7) 94,021,051,000
8) 58,000,708,120
9) 59,000,054,086
10) 74,000,354,158
11) 7,013,012,007
12) 72,450,000,658
13) 398,067,128
14) 24,054,009,032
15) 795,000,458
16) 38,002,054,009

Numbers in Word Form

1) three hundred seventy-two
2) six hundred five
3) five hundred fifty
4) three hundred fifty-one
5) seven hundred ninety-three
6) six hundred forty-seven
7) three thousand, two hundred nineteen
8) five thousand, three hundred twenty-six
9) two thousand, eight hundred forty-two
10) four thousand, six hundred ninety-one
11) five thousand, five hundred thirty-one
12) seven thousand, three hundred sixty
13) two thousand, five hundred thirty-two
14) eight thousand, fourteen

STAAR Subject Test Mathematics Grade 4

15) eleven thousand, two hundred forty-two

Roman Numerals

1) XI
2) XXI
3) XXIV
4) XVI
5) XXVII
6) XXIX
7) XII
8) XXVIII
9) XV
10) XX
11) XXX
12) XXIX

Rounding Numbers

1) 4,000
2) 4,000
3) 3,450
4) 7,190
5) 5,280
6) 1,480
7) 8,300
8) 24.70
9) 85.00
10) 71.30
11) 665.00
12) 9,000
13) 15.380
14) 4,500
15) 37.00
16) 4,800
17) 6,700
18) 2,540
19) 73.60
20) 16,500
21) 30.00
22) 12,370
23) 31,700
24) 7,840

Odd or Even

1) even
2) odd
3) odd
4) odd
5) odd
6) even
7) even
8) odd
9) even
10) even
11) odd
12) odd
13) 52
14) 18
15) 86
16) 32
17) 71
18) 67
19) 25
20) 89

STAAR Subject Test Mathematics Grade 4

Chapter 2 : Whole Number Operations

Topics that you'll learn in this chapter:

- ✓ Adding Whole Numbers,
- ✓ Subtracting Whole Numbers,
- ✓ Multiplying Whole Numbers,
- ✓ Dividing Hundreds,
- ✓ Long Division by One Digit,
- ✓ Division with Remainders,
- ✓ Rounding Whole Numbers,
- ✓ Whole Number Estimation,

STAAR Subject Test Mathematics Grade 4

Adding Whole Numbers

✏ Add.

1) 5,763
   + 8,238
   _____

2) 6,834
   + 4,998
   _____

3) 3,548
   + 5,693
   _____

4) 2,769
   + 8,872
   _____

5) 3,196
   + 2,936
   _____

6) 7,009
   + 4,992
   _____

✏ Find the missing numbers.

7) 3,468 + ___ = 4,102

8) 840 + 2,360 = ___

9) 5,200 + ___ = 7,980

10) 631 + ___ = 2,007

11) ___ + 803 = 3,945

12) ___ + 2,156 = 5,922

13) David sells gems. He finds a diamond in Istanbul and buys it for $4,795. Then, he flies to Cairo and purchases a bigger diamond for the bargain price of $9,633. How much does David spend on the two diamonds? _____

WWW.MathNotion.Com

STAAR Subject Test Mathematics Grade 4

Subtracting Whole Numbers

✎ Subtract.

1) 10,512
   − 4,411
   _____

2) 5,204
   − 3,679
   _____

3) 8,520
   − 6,483
   _____

4) 8,001
   − 5,224
   _____

5) 11,916
   − 8,711
   _____

6) 5,005
   − 2,008
   _____

✎ Find the missing number.

7) 5,263 − ___ = 2,367

8) 7,198 − ___ = 4,742

9) 8,928 − 3,764 = ___

10) 6,511 − ___ = 3,759

11) 7,003 − 5,489 = ___

12) 8,800 − 5,995 = ___

13) Jackson had $7,189 invested in the stock market until he lost $3,793 on those investments. How much money does he have in the stock market now?

_____

STAAR Subject Test Mathematics Grade 4

Multiplying Whole Numbers

✎ Find the answers.

1) 2,200
   × 31
   ─────

2) 3,200
   × 22
   ─────

3) 5,790
   × 5
   ─────

4) 5,220
   × 3
   ─────

5) 6,911
   × 3
   ─────

6) 1,998
   ×40
   ─────

7) 2,893
   ×5.5
   ─────

8) 2,254
   × 3.5
   ─────

9) 4,372
   × 4.8
   ─────

10) 3,984
    × 2.75
    ─────

11) 4,900
    × 2.5
    ─────

12) 8,200
    × 4.5
    ─────

WWW.MathNotion.Com   24

STAAR Subject Test Mathematics Grade 4

Dividing Hundreds

✏ Find answers.

1) 4,440 ÷ 400

2) 1,600 ÷ 40

3) 9,990 ÷ 90

4) 4,200 ÷ 60

5) 6,400 ÷ 8,000

6) 2,700 ÷ 30

7) 3,333 ÷ 30

8) 558 ÷ 45

9) 2,278 ÷ 85

10) 1,683 ÷ 55

11) 1,582 ÷ 35

12) 9,000 ÷ 600

13) 1,000 ÷ 2,500

14) 44.8 ÷ 20

15) 6,800 ÷ 400

16) 1,500 ÷ 5,000

17) 36.60 ÷ 120

18) 7,700 ÷ 700

19) 5,400 ÷ 600

20) 8,000 ÷ 160

21) 18,000 ÷ 9,000

22) 42,000 ÷ 30

23) 480 ÷ 40

24) 63,000 ÷ 900

STAAR Subject Test Mathematics Grade 4

Long Division by Two Digits

✎ Find the quotient.

1) 18)576

2) 14)952

3) 21)588

4) 23)299

5) 44)748

6) 26)234

7) 16)496

8) 29)1,479

9) 54)1,080

10) 41)1,476

11) 53)2,491

12) 60)2,880

13) 32)2,912

14) 77)8,393

15) 85)3,740

16) 57)4,617

17) 50)9,200

18) 25)15,400

Division with Remainders

✎ Find the quotient with remainder.

1) 14)715

2) 16)2,750

3) 27)4,603

4) 58)2,554

5) 42)7,732

6) 63)6,737

7) 71)9,036

8) 65)8,624

9) 35)5,705

10) 92)13,161

11) 46)12,214

12) 69)42,482

13) 85)6,858

14) 87)34,304

STAAR Subject Test Mathematics Grade 4

Rounding Whole Numbers

✎ Round each number to the underlined place value.

1) 7,5̲33

2) 9,3̲74

3) 8,8̲83

4) 2,3̲68

5) 5,5̲77

6) 3,3̲81

7) 3,5̲20

8) 9,3̲38

9) 8.5̲81

10) 33.5̲7

11) 51.6̲9

12) 22.1̲38

13) 6̲,758

14) 11,5̲57

15) 8,8̲38

16) 5.8̲89

17) 1.8̲60

18) 25.0̲70

19) 9̲.332

20) 49.4̲8

21) 28.8̲9

22) 24,3̲77

23) 52,1̲58

24) 13,8̲83

25) 9,6̲09

26) 17,4̲51

27) 18,7̲68

www.MathNotion.Com

STAAR Subject Test Mathematics Grade 4

Whole Number Estimation

✎ Estimate the sum by rounding each added to the nearest ten.

1) 875 + 325

2) 985 + 1,452

3) 2,424 + 4,128

4) 1,576 + 6,279

5) 1,247 + 3,863

6) 6,746 + 5,121

7) 3,924 + 6,456

8) 1,785 + 7,164

9) $\begin{array}{r} 1,458 \\ + 2,442 \\ \hline \end{array}$

10) $\begin{array}{r} 5,689 \\ + 4,151 \\ \hline \end{array}$

11) $\begin{array}{r} 8,259 \\ + 4,754 \\ \hline \end{array}$

12) $\begin{array}{r} 6,788 \\ + 3,954 \\ \hline \end{array}$

13) $\begin{array}{r} 9,123 \\ + 4,455 \\ \hline \end{array}$

14) $\begin{array}{r} 6,680 \\ + 5,358 \\ \hline \end{array}$

15) $\begin{array}{r} 3,165 \\ + 7,124 \\ \hline \end{array}$

16) $\begin{array}{r} 8,859 \\ + 6,452 \\ \hline \end{array}$

STAAR Subject Test Mathematics Grade 4

Answers of Worksheets

Adding Whole Numbers

1) 14,001
2) 11,832
3) 9,241
4) 11,641
5) 6,132
6) 12,001
7) 634
8) 3,200
9) 2,780
10) 1,376
11) 3,142
12) 3,766
13) $14,428

Subtracting Whole Numbers

1) 6,101
2) 1,525
3) 2,037
4) 2,777
5) 3,205
6) 2,997
7) 2,896
8) 2,456
9) 5,164
10) 2,752
11) 1,514
12) 2,805
13) 3,396

Multiplying Whole Numbers

1) 68,200
2) 70,400
3) 28,950
4) 15,660
5) 20,733
6) 79,920
7) 15,911.5
8) 7,889
9) 20,985.6
10) 10,956
11) 12,250
12) 36,900

Dividing Hundreds

1) 11.1
2) 40
3) 111
4) 70
5) 0.8
6) 90
7) 111.1
8) 12.4
9) 26.8
10) 30.6
11) 45.2
12) 15
13) 0.4
14) 2.24
15) 17
16) 0.3
17) 0.305
18) 11
19) 9
20) 50
21) 2
22) 1,400
23) 12
24) 70

Long Division by Two Digits

1) 32
2) 68
3) 28
4) 13
5) 17
6) 9
7) 31
8) 51
9) 20
10) 36
11) 47
12) 48
13) 91
14) 109
15) 44
16) 81
17) 184
18) 616

WWW.MathNotion.Com

STAAR Subject Test Mathematics Grade 4

Division with Remainders

1) 51 R1
2) 171 R14
3) 170 R13
4) 44 R2
5) 184 R4
6) 106 R59
7) 127 R19
8) 132 R44
9) 163 R0
10) 143 R5
11) 265 R24
12) 615 R47
13) 80 R58
14) 394 R26

Rounding Whole Numbers

1) 7,500
2) 9,400
3) 8,880
4) 2,370
5) 5,580
6) 3,380
7) 3,500
8) 9,340
9) 8.60
10) 33.60
11) 51.70
12) 22.100
13) 7,000
14) 11,560
15) 8,840
16) 5.900
17) 1.900
18) 25.100
19) 9.000
20) 49.50
21) 28.90
22) 24,380
23) 52,160
24) 13,880
25) 9,600
26) 17,450
27) 18,800

Whole Number Estimation

1) 1,200
2) 2,440
3) 6,550
4) 7,860
5) 5,110
6) 11,870
7) 10,380
8) 8,950
9) 3,900
10) 9,840
11) 13,010
12) 10,740
13) 13,580
14) 12,040
15) 10,290
16) 15,310

Chapter 3 : Fractions

Topics that you'll learn in this chapter:

- ✓ Simplifying Fractions,
- ✓ Like Denominators,
- ✓ Compare Fractions with Like Denominators,
- ✓ More than two Fractions with Like Denominators,
- ✓ Unlike Denominators,
- ✓ Ordering Fractions,
- ✓ Denominators of 10, 100, and 1000,
- ✓ Multiply Fraction

STAAR Subject Test Mathematics Grade 4

Simplifying Fractions

✎ Simplify the fractions.

1) $\dfrac{44}{84}$

2) $\dfrac{8}{20}$

3) $\dfrac{12}{16}$

4) $\dfrac{4}{24}$

5) $\dfrac{15}{30}$

6) $\dfrac{9}{63}$

7) $\dfrac{4}{14}$

8) $\dfrac{17}{51}$

9) $\dfrac{24}{30}$

10) $\dfrac{5}{35}$

11) $\dfrac{16}{48}$

12) $\dfrac{33}{22}$

13) $\dfrac{45}{63}$

14) $\dfrac{2.4}{3.2}$

15) $\dfrac{12}{60}$

16) $\dfrac{70}{112}$

17) $\dfrac{2.7}{7.2}$

18) $\dfrac{33}{88}$

19) $\dfrac{1.5}{13.5}$

20) $\dfrac{39}{52}$

21) $\dfrac{5}{45}$

22) $\dfrac{2.1}{4.2}$

WWW.MathNotion.Com

Like Denominators

✏️ Add fractions.

1) $\dfrac{3}{4}+\dfrac{1}{4}$

2) $\dfrac{1}{5}+\dfrac{4}{5}$

3) $\dfrac{4}{9}+\dfrac{7}{9}$

4) $\dfrac{2}{7}+\dfrac{2}{7}$

5) $\dfrac{5}{13}+\dfrac{2}{13}$

6) $\dfrac{1}{14}+\dfrac{4}{14}$

7) $\dfrac{11}{19}+\dfrac{1}{19}$

8) $\dfrac{3}{16}+\dfrac{9}{16}$

9) $\dfrac{3}{10}+\dfrac{1}{10}$

10) $\dfrac{6}{17}+\dfrac{2}{17}$

11) $\dfrac{5}{22}+\dfrac{5}{22}$

12) $\dfrac{7}{35}+\dfrac{11}{35}$

13) $\dfrac{7}{27}+\dfrac{20}{27}$

14) $\dfrac{2}{31}+\dfrac{10}{31}$

15) $\dfrac{5}{23}+\dfrac{3}{23}$

16) $\dfrac{8}{41}+\dfrac{13}{41}$

17) $\dfrac{15}{37}+\dfrac{18}{37}$

18) $\dfrac{2}{51}+\dfrac{7}{51}$

19) $\dfrac{17}{26}+\dfrac{6}{26}$

20) $\dfrac{12}{48}+\dfrac{11}{48}$

21) $\dfrac{11}{29}+\dfrac{8}{29}$

22) $\dfrac{15}{34}+\dfrac{19}{34}$

23) $\dfrac{1}{19}+\dfrac{5}{19}$

24) $\dfrac{3}{53}+\dfrac{4}{53}$

25) $\dfrac{3}{20}+\dfrac{6}{20}$

26) $\dfrac{2}{63}+\dfrac{6}{63}$

27) $\dfrac{6}{38}+\dfrac{1}{38}$

28) $\dfrac{14}{31}+\dfrac{17}{31}$

29) $\dfrac{3}{28}+\dfrac{5}{28}$

30) $\dfrac{2}{37}+\dfrac{15}{37}$

STAAR Subject Test Mathematics Grade 4

🖎 **Subtract fractions.**

1) $\dfrac{8}{9} - \dfrac{4}{9}$

2) $\dfrac{3}{8} - \dfrac{1}{8}$

3) $\dfrac{9}{11} - \dfrac{3}{11}$

4) $\dfrac{9}{14} - \dfrac{4}{14}$

5) $\dfrac{15}{20} - \dfrac{8}{20}$

6) $\dfrac{8}{15} - \dfrac{7}{15}$

7) $\dfrac{11}{19} - \dfrac{9}{19}$

8) $\dfrac{13}{16} - \dfrac{1}{16}$

9) $\dfrac{7}{29} - \dfrac{4}{29}$

10) $\dfrac{14}{23} - \dfrac{7}{23}$

11) $\dfrac{15}{34} - \dfrac{7}{34}$

12) $\dfrac{18}{41} - \dfrac{9}{41}$

13) $\dfrac{17}{39} - \dfrac{16}{39}$

14) $\dfrac{6}{26} - \dfrac{2}{26}$

15) $\dfrac{14}{17} - \dfrac{4}{17}$

16) $\dfrac{33}{55} - \dfrac{20}{55}$

17) $\dfrac{41}{49} - \dfrac{36}{49}$

18) $\dfrac{40}{53} - \dfrac{39}{53}$

19) $\dfrac{27}{37} - \dfrac{17}{37}$

20) $\dfrac{21}{47} - \dfrac{11}{47}$

21) $\dfrac{24}{43} - \dfrac{12}{43}$

22) $\dfrac{13}{19} - \dfrac{12}{19}$

23) $\dfrac{6}{26} - \dfrac{3}{26}$

24) $\dfrac{9}{15} - \dfrac{7}{15}$

25) $\dfrac{8}{39} - \dfrac{3}{39}$

26) $\dfrac{18}{61} - \dfrac{15}{61}$

27) $\dfrac{12}{53} - \dfrac{9}{53}$

28) $\dfrac{75}{76} - \dfrac{74}{76}$

29) $\dfrac{26}{45} - \dfrac{13}{45}$

30) $\dfrac{20}{57} - \dfrac{17}{57}$

STAAR Subject Test Mathematics Grade 4

Compare Fractions with Like Denominators

✎ Evaluate and compare. Write < or > or =.

1) $\frac{1}{3} + \frac{1}{3} \underline{\phantom{<}} \frac{1}{3}$

2) $\frac{3}{6} + \frac{3}{6} \underline{\phantom{<}} \frac{5}{6}$

3) $\frac{8}{9} - \frac{4}{9} \underline{\phantom{<}} \frac{7}{9}$

4) $\frac{4}{11} + \frac{5}{11} \underline{\phantom{<}} \frac{7}{11}$

5) $\frac{9}{14} - \frac{8}{14} \underline{\phantom{<}} \frac{5}{14}$

6) $\frac{11}{17} - \frac{3}{17} \underline{\phantom{<}} \frac{6}{17}$

7) $\frac{11}{21} + \frac{2}{21} \underline{\phantom{<}} \frac{10}{21}$

8) $\frac{8}{32} + \frac{6}{32} \underline{\phantom{<}} \frac{9}{32}$

9) $\frac{25}{29} - \frac{16}{29} \underline{\phantom{<}} \frac{11}{29}$

10) $\frac{28}{41} + \frac{13}{41} \underline{\phantom{<}} \frac{27}{41}$

11) $\frac{18}{35} - \frac{11}{35} \underline{\phantom{<}} \frac{22}{35}$

12) $\frac{32}{47} - \frac{22}{47} \underline{\phantom{<}} \frac{11}{47}$

13) $\frac{14}{27} + \frac{13}{27} \underline{\phantom{<}} \frac{24}{27}$

14) $\frac{34}{52} - \frac{11}{52} \underline{\phantom{<}} \frac{21}{52}$

15) $\frac{43}{56} - \frac{24}{56} \underline{\phantom{<}} \frac{27}{56}$

16) $\frac{27}{71} + \frac{25}{71} \underline{\phantom{<}} \frac{48}{71}$

WWW.MathNotion.Com

STAAR Subject Test Mathematics Grade 4

More Than Two Fractions with Like Denominators

✎ Add fractions.

1) $\dfrac{5}{9} + \dfrac{2}{9} + \dfrac{2}{9}$

2) $\dfrac{4}{6} + \dfrac{1}{6} + \dfrac{1}{6}$

3) $\dfrac{2}{17} + \dfrac{4}{17} + \dfrac{2}{17}$

4) $\dfrac{1}{5} + \dfrac{1}{5} + \dfrac{1}{5}$

5) $\dfrac{7}{18} + \dfrac{2}{18} + \dfrac{3}{18}$

6) $\dfrac{3}{27} + \dfrac{5}{27} + \dfrac{2}{27}$

7) $\dfrac{4}{33} + \dfrac{4}{33} + \dfrac{4}{33}$

8) $\dfrac{8}{23} + \dfrac{6}{23} + \dfrac{2}{23}$

9) $\dfrac{13}{41} + \dfrac{2}{41} + \dfrac{8}{41}$

10) $\dfrac{6}{35} + \dfrac{9}{35} + \dfrac{20}{35}$

11) $\dfrac{1}{37} + \dfrac{5}{37} + \dfrac{5}{37}$

12) $\dfrac{4}{43} + \dfrac{9}{43} + \dfrac{8}{43}$

13) $\dfrac{4}{51} + \dfrac{10}{51} + \dfrac{7}{51}$

14) $\dfrac{5}{26} + \dfrac{13}{26} + \dfrac{6}{26}$

15) $\dfrac{5}{64} + \dfrac{4}{64} + \dfrac{2}{64}$

16) $\dfrac{1}{73} + \dfrac{5}{73} + \dfrac{6}{73}$

STAAR Subject Test Mathematics Grade 4

Unlike Denominators

✎ Add fraction.

1) $\dfrac{2}{9} + \dfrac{3}{4}$

2) $\dfrac{1}{4} + \dfrac{3}{5}$

3) $\dfrac{1}{16} + \dfrac{3}{4}$

4) $\dfrac{3}{8} + \dfrac{1}{7}$

5) $\dfrac{1}{3} + \dfrac{2}{4}$

6) $\dfrac{1}{6} + \dfrac{3}{7}$

7) $\dfrac{5}{18} + \dfrac{4}{6}$

8) $\dfrac{1}{12} + \dfrac{5}{6}$

9) $\dfrac{5}{27} + \dfrac{1}{9}$

10) $\dfrac{1}{6} + \dfrac{7}{24}$

11) $\dfrac{3}{5} + \dfrac{1}{8}$

12) $\dfrac{11}{42} + \dfrac{3}{7}$

13) $\dfrac{7}{20} + \dfrac{1}{3}$

14) $\dfrac{1}{45} + \dfrac{3}{5}$

15) $\dfrac{3}{32} + \dfrac{5}{8}$

16) $\dfrac{3}{48} + \dfrac{5}{6}$

17) $\dfrac{5}{12} + \dfrac{1}{6}$

18) $\dfrac{1}{34} + \dfrac{3}{17}$

19) $\dfrac{4}{9} + \dfrac{7}{54}$

20) $\dfrac{13}{56} + \dfrac{4}{7}$

21) $\dfrac{3}{12} + \dfrac{2}{3}$

22) $\dfrac{4}{33} + \dfrac{5}{11}$

WWW.MathNotion.Com

STAAR Subject Test Mathematics Grade 4

✎ Subtract fractions.

1) $\dfrac{8}{9} - \dfrac{1}{2}$

2) $\dfrac{2}{3} - \dfrac{3}{10}$

3) $\dfrac{1}{6} - \dfrac{1}{9}$

4) $\dfrac{7}{8} - \dfrac{1}{4}$

5) $\dfrac{3}{4} - \dfrac{1}{28}$

6) $\dfrac{11}{30} - \dfrac{3}{15}$

7) $\dfrac{11}{18} - \dfrac{5}{9}$

8) $\dfrac{5}{13} - \dfrac{3}{26}$

9) $\dfrac{17}{35} - \dfrac{2}{7}$

10) $\dfrac{5}{6} - \dfrac{12}{36}$

11) $\dfrac{5}{9} - \dfrac{1}{27}$

12) $\dfrac{3}{5} - \dfrac{1}{8}$

13) $\dfrac{2}{3} - \dfrac{3}{5}$

14) $\dfrac{7}{8} - \dfrac{3}{7}$

15) $\dfrac{5}{9} - \dfrac{13}{45}$

16) $\dfrac{3}{4} - \dfrac{5}{36}$

17) $\dfrac{39}{49} - \dfrac{5}{7}$

18) $\dfrac{3}{11} - \dfrac{3}{22}$

19) $\dfrac{17}{48} - \dfrac{4}{12}$

20) $\dfrac{2}{3} - \dfrac{4}{13}$

21) $\dfrac{5}{8} - \dfrac{19}{72}$

22) $\dfrac{3}{5} - \dfrac{1}{12}$

WWW.MathNotion.Com

STAAR Subject Test Mathematics Grade 4

Ordering Fractions

✏ Order the fractions from least to greatest.

1) $\frac{1}{5}, \frac{1}{11}, \frac{1}{8}, \frac{1}{3}$ _____, _____, _____, _____

2) $\frac{1}{9}, \frac{1}{18}, \frac{2}{4}, \frac{1}{5}$ _____, _____, _____, _____

3) $\frac{4}{7}, \frac{1}{7}, \frac{6}{21}, \frac{15}{21}$ _____, _____, _____, _____

4) $\frac{1}{2}, \frac{1}{3}, \frac{4}{9}, \frac{5}{18}$ _____, _____, _____, _____

5) $\frac{4}{9}, \frac{3}{4}, \frac{7}{36}, \frac{1}{6}$ _____, _____, _____, _____

✏ Order the fractions from greatest to least.

6) $\frac{3}{4}, \frac{4}{7}, \frac{3}{10}, \frac{5}{13}$ _____, _____, _____, _____

7) $\frac{5}{11}, \frac{5}{6}, \frac{2}{5}, \frac{1}{3}$ _____, _____, _____, _____

8) $\frac{7}{8}, \frac{1}{6}, \frac{3}{4}, \frac{5}{15}$ _____, _____, _____, _____

9) $\frac{4}{7}, \frac{2}{3}, \frac{11}{25}, \frac{13}{33}$ _____, _____, _____, _____

10) $\frac{18}{20}, \frac{15}{16}, \frac{14}{18}, \frac{5}{12}$ _____, _____, _____, _____

STAAR Subject Test Mathematics Grade 4

Denominators of 10, 100, and 1000

✎ Add fractions.

1) $\dfrac{7}{10} + \dfrac{13}{100}$

2) $\dfrac{1}{10} + \dfrac{10}{100}$

3) $\dfrac{15}{100} + \dfrac{1}{1,000}$

4) $\dfrac{56}{100} + \dfrac{3}{10}$

5) $\dfrac{50}{1,000} + \dfrac{7}{10}$

6) $\dfrac{6}{10} + \dfrac{30}{1,000}$

7) $\dfrac{9}{100} + \dfrac{3}{10}$

8) $\dfrac{5}{10} + \dfrac{50}{100}$

9) $\dfrac{48}{100} + \dfrac{6}{10}$

10) $\dfrac{70}{100} + \dfrac{2}{10}$

11) $\dfrac{80}{100} + \dfrac{200}{1,000}$

12) $\dfrac{30}{100} + \dfrac{4}{10}$

13) $\dfrac{9}{100} + \dfrac{7}{10}$

14) $\dfrac{25}{100} + \dfrac{6}{10}$

15) $\dfrac{15}{100} + \dfrac{8}{10}$

16) $\dfrac{3}{10} + \dfrac{31}{100}$

17) $\dfrac{8}{10} + \dfrac{11}{100}$

18) $\dfrac{34}{100} + \dfrac{6}{10}$

WWW.MathNotion.Com

STAAR Subject Test Mathematics Grade 4

✏️ Subtract fractions.

1) $\dfrac{8}{10} - \dfrac{20}{100}$

2) $\dfrac{5}{10} - \dfrac{47}{100}$

3) $\dfrac{12}{100} - \dfrac{60}{1,000}$

4) $\dfrac{6}{10} - \dfrac{50}{100}$

5) $\dfrac{3}{10} - \dfrac{23}{100}$

6) $\dfrac{70}{100} - \dfrac{250}{1,000}$

7) $\dfrac{4}{10} - \dfrac{350}{1,000}$

8) $\dfrac{70}{100} - \dfrac{3}{10}$

9) $\dfrac{40}{100} - \dfrac{3}{10}$

10) $\dfrac{6}{10} - \dfrac{180}{1,000}$

11) $\dfrac{93}{100} - \dfrac{5}{10}$

12) $\dfrac{65}{100} - \dfrac{4}{10}$

13) $\dfrac{80}{100} - \dfrac{6}{10}$

14) $\dfrac{90}{100} - \dfrac{5}{10}$

15) $\dfrac{200}{1,000} - \dfrac{1}{10}$

16) $\dfrac{90}{100} - \dfrac{7}{10}$

17) $\dfrac{900}{1,000} - \dfrac{40}{100}$

18) $\dfrac{60}{100} - \dfrac{3}{10}$

STAAR Subject Test Mathematics Grade 4

Multiplying Fractions

Find the product.

1) $\dfrac{4}{5} \times \dfrac{2}{6} =$

2) $\dfrac{4}{22} \times \dfrac{5}{8} =$

3) $\dfrac{8}{30} \times \dfrac{12}{16} =$

4) $\dfrac{9}{14} \times \dfrac{21}{36} =$

5) $\dfrac{14}{15} \times \dfrac{5}{7} =$

6) $\dfrac{16}{19} \times \dfrac{3}{4} =$

7) $\dfrac{4}{9} \times \dfrac{9}{8} =$

8) $\dfrac{47}{85} \times 0 =$

9) $\dfrac{5}{8} \times \dfrac{16}{6} =$

10) $\dfrac{28}{15} \times \dfrac{5}{7} =$

11) $\dfrac{32}{24} \times \dfrac{12}{16} =$

12) $\dfrac{6}{42} \times \dfrac{7}{36} =$

13) $\dfrac{13}{8} \times \dfrac{12}{4} =$

14) $\dfrac{10}{9} \times \dfrac{6}{5} =$

15) $\dfrac{35}{56} \times \dfrac{8}{7} =$

16) $\dfrac{16}{18} \times 9 =$

17) $\dfrac{5}{22} \times \dfrac{44}{15} =$

18) $\dfrac{10}{18} \times \dfrac{9}{20} =$

19) $\dfrac{7}{11} \times \dfrac{8}{21} =$

20) $\dfrac{26}{24} \times \dfrac{8}{52} =$

21) $\dfrac{6}{17} \times \dfrac{1}{12} =$

22) $\dfrac{20}{9} \times \dfrac{6}{100} =$

23) $\dfrac{8}{14} \times \dfrac{7}{72} =$

24) $\dfrac{50}{100} \times \dfrac{300}{400} =$

WWW.MathNotion.Com

STAAR Subject Test Mathematics Grade 4

Answers of Worksheets

Simplifying Fractions

1) $\frac{11}{21}$
2) $\frac{2}{5}$
3) $\frac{3}{4}$
4) $\frac{1}{6}$
5) $\frac{1}{2}$
6) $\frac{1}{7}$
7) $\frac{2}{7}$
8) $\frac{1}{3}$
9) $\frac{4}{5}$
10) $\frac{1}{7}$
11) $\frac{1}{3}$
12) $\frac{3}{2}$
13) $\frac{5}{7}$
14) $\frac{3}{4}$
15) $\frac{1}{5}$
16) $\frac{5}{8}$
17) $\frac{3}{8}$
18) $\frac{3}{8}$
19) $\frac{1}{9}$
20) $\frac{3}{4}$
21) $\frac{1}{9}$
22) $\frac{1}{2}$

Like Denominators (addition)

1) 1
2) 1
3) $\frac{11}{9}$
4) $\frac{4}{7}$
5) $\frac{7}{13}$
6) $\frac{5}{14}$
7) $\frac{12}{19}$
8) $\frac{3}{4}$
9) $\frac{2}{5}$
10) $\frac{8}{17}$
11) $\frac{5}{11}$
12) $\frac{18}{35}$
13) 1
14) $\frac{12}{31}$
15) $\frac{8}{23}$
16) $\frac{21}{41}$
17) $\frac{33}{37}$
18) $\frac{3}{17}$
19) $\frac{23}{26}$
20) $\frac{23}{48}$
21) $\frac{19}{29}$
22) 1
23) $\frac{6}{19}$
24) $\frac{7}{53}$
25) $\frac{9}{20}$
26) $\frac{8}{63}$
27) $\frac{7}{38}$
28) 1
29) $\frac{2}{7}$
30) $\frac{17}{37}$

Like Denominators (Subtraction)

1) $\frac{4}{9}$
2) $\frac{1}{4}$
3) $\frac{6}{11}$
4) $\frac{5}{14}$
5) $\frac{7}{20}$
6) $\frac{1}{15}$
7) $\frac{2}{19}$
8) $\frac{3}{4}$
9) $\frac{3}{29}$
10) $\frac{7}{23}$
11) $\frac{8}{34}$
12) $\frac{9}{41}$
13) $\frac{1}{39}$
14) $\frac{2}{13}$
15) $\frac{10}{17}$
16) $\frac{13}{55}$
17) $\frac{5}{49}$
18) $\frac{1}{53}$
19) $\frac{10}{37}$
20) $\frac{10}{47}$
21) $\frac{12}{43}$
22) $\frac{1}{19}$
23) $\frac{3}{26}$
24) $\frac{2}{15}$
25) $\frac{5}{39}$
26) $\frac{3}{61}$
27) $\frac{3}{53}$
28) $\frac{1}{76}$

WWW.MathNotion.Com

STAAR Subject Test Mathematics Grade 4

29) $\frac{13}{45}$ 30) $\frac{1}{19}$

Compare Fractions with Like Denominators

1) $\frac{2}{3} > \frac{1}{3}$ 5) $\frac{1}{14} < \frac{5}{14}$ 9) $\frac{9}{29} < \frac{11}{29}$ 13) $1 > \frac{24}{27}$

2) $1 > \frac{5}{6}$ 6) $\frac{8}{17} > \frac{6}{17}$ 10) $1 > \frac{27}{41}$ 14) $\frac{23}{52} > \frac{21}{52}$

3) $\frac{4}{9} < \frac{7}{9}$ 7) $\frac{13}{21} > \frac{10}{21}$ 11) $\frac{7}{35} < \frac{22}{35}$ 15) $\frac{19}{56} < \frac{27}{56}$

4) $\frac{9}{11} > \frac{7}{11}$ 8) $\frac{14}{32} > \frac{9}{32}$ 12) $\frac{10}{47} < \frac{11}{47}$ 16) $\frac{52}{71} > \frac{48}{71}$

More Than Two Fractions with Like Denominators

1) 1   5) $\frac{2}{3}$ 9) $\frac{23}{41}$ 13) $\frac{7}{17}$

2) 1   6) $\frac{10}{27}$ 10) 1   14) $\frac{12}{13}$

3) $\frac{8}{17}$ 7) $\frac{4}{11}$ 11) $\frac{11}{37}$ 15) $\frac{11}{64}$

4) $\frac{3}{5}$ 8) $\frac{16}{23}$ 12) $\frac{21}{43}$ 16) $\frac{12}{73}$

Unlike Denominators (Addition)

1) $\frac{35}{36}$ 7) $\frac{17}{18}$ 13) $\frac{41}{60}$ 19) $\frac{31}{54}$

2) $\frac{17}{20}$ 8) $\frac{11}{12}$ 14) $\frac{28}{45}$ 20) $\frac{45}{56}$

3) $\frac{13}{16}$ 9) $\frac{8}{27}$ 15) $\frac{23}{32}$ 21) $\frac{11}{12}$

4) $\frac{29}{56}$ 10) $\frac{11}{24}$ 16) $\frac{43}{48}$ 22) $\frac{19}{33}$

5) $\frac{5}{6}$ 11) $\frac{29}{40}$ 17) $\frac{7}{12}$

6) $\frac{25}{42}$ 12) $\frac{29}{42}$ 18) $\frac{7}{34}$

Unlike Denominators (Subtraction)

1) $\frac{7}{18}$ 7) $\frac{1}{18}$ 13) $\frac{1}{15}$ 19) $\frac{1}{48}$

2) $\frac{11}{30}$ 8) $\frac{7}{26}$ 14) $\frac{25}{56}$ 20) $\frac{14}{39}$

3) $\frac{1}{18}$ 9) $\frac{1}{5}$ 15) $\frac{4}{15}$ 21) $\frac{13}{36}$

4) $\frac{5}{8}$ 10) $\frac{1}{2}$ 16) $\frac{11}{18}$ 22) $\frac{31}{60}$

5) $\frac{5}{7}$ 11) $\frac{14}{27}$ 17) $\frac{4}{49}$

6) $\frac{1}{6}$ 12) $\frac{19}{40}$ 18) $\frac{3}{22}$

WWW.MathNotion.Com

STAAR Subject Test Mathematics Grade 4

Ordering Fractions

1) $\frac{1}{11}, \frac{1}{8}, \frac{1}{5}, \frac{1}{3}$

2) $\frac{1}{18}, \frac{1}{9}, \frac{1}{5}, \frac{2}{4}$

3) $\frac{1}{7}, \frac{6}{21}, \frac{4}{7}, \frac{15}{21}$

4) $\frac{5}{18}, \frac{1}{3}, \frac{4}{9}, \frac{1}{2}$

5) $\frac{1}{6}, \frac{7}{36}, \frac{4}{9}, \frac{3}{4}$

6) $\frac{3}{4}, \frac{4}{7}, \frac{5}{13}, \frac{3}{10}$

7) $\frac{5}{6}, \frac{5}{11}, \frac{2}{5}, \frac{1}{3}$

8) $\frac{7}{8}, \frac{3}{4}, \frac{5}{15}, \frac{1}{6}$

9) $\frac{2}{3}, \frac{4}{7}, \frac{11}{25}, \frac{13}{33}$

10) $\frac{15}{16}, \frac{18}{20}, \frac{14}{18}, \frac{5}{12}$

Denominators of 10, 100, and 1000

1) $\frac{83}{100}$
2) $\frac{1}{5}$
3) $\frac{151}{1,000}$
4) $\frac{43}{50}$
5) $\frac{3}{4}$
6) $\frac{63}{100}$
7) $\frac{39}{100}$
8) 1
9) $\frac{27}{25}$
10) $\frac{9}{10}$
11) 1
12) $\frac{7}{10}$
13) $\frac{79}{100}$
14) $\frac{17}{20}$
15) $\frac{19}{20}$
16) $\frac{61}{100}$
17) $\frac{91}{100}$
18) $\frac{47}{50}$

Denominators of 10, 100, and 1000 (Subtract)

1) $\frac{3}{5}$
2) $\frac{3}{100}$
3) $\frac{3}{50}$
4) $\frac{1}{10}$
5) $\frac{7}{100}$
6) $\frac{9}{20}$
7) $\frac{1}{20}$
8) $\frac{2}{5}$
9) $\frac{1}{10}$
10) $\frac{21}{50}$
11) $\frac{43}{100}$
12) $\frac{1}{4}$
13) $\frac{1}{5}$
14) $\frac{2}{5}$
15) $\frac{1}{10}$
16) $\frac{1}{5}$
17) $\frac{1}{2}$
18) $\frac{3}{10}$

Multiplying Fractions

1) $\frac{4}{15}$
2) $\frac{5}{44}$
3) $\frac{1}{5}$
4) $\frac{3}{8}$
5) $\frac{2}{3}$
6) $\frac{12}{19}$
7) $\frac{1}{2}$
8) 0
9) $\frac{5}{3}$
10) $\frac{4}{3}$
11) 1
12) $\frac{1}{36}$
13) $\frac{39}{8}$
14) $\frac{4}{3}$
15) $\frac{5}{7}$
16) 8
17) $\frac{2}{3}$
18) $\frac{1}{4}$
19) $\frac{8}{33}$
20) $\frac{1}{6}$
21) $\frac{1}{34}$
22) $\frac{2}{15}$
23) $\frac{1}{18}$
24) $\frac{3}{8}$

Chapter 4:
Mixed Numbers

Topics that you'll learn in this chapter:

- ✓ Fractions to Mixed Numbers,
- ✓ Mixed Numbers to Fractions,
- ✓ Add and Subtract Mixed Numbers,
- ✓ Multiply Mixed Numbers

Fractions to Mixed Numbers

✎ Convert fractions to mixed numbers.

1) $\dfrac{9}{5}$

2) $\dfrac{11}{3}$

3) $\dfrac{39}{8}$

4) $\dfrac{27}{11}$

5) $\dfrac{7}{2}$

6) $\dfrac{43}{4}$

7) $\dfrac{49}{9}$

8) $\dfrac{15}{4}$

9) $\dfrac{37}{7}$

10) $\dfrac{19}{7}$

11) $\dfrac{41}{9}$

12) $\dfrac{45}{12}$

13) $\dfrac{17}{5}$

14) $\dfrac{29}{6}$

15) $\dfrac{13}{4}$

16) $\dfrac{15}{7}$

17) $\dfrac{65}{7}$

18) $\dfrac{59}{8}$

19) $\dfrac{25}{4}$

20) $\dfrac{17}{8}$

STAAR Subject Test Mathematics Grade 4

Mixed Numbers to Fractions

Convert to fraction.

1) $3\frac{3}{5}$

2) $1\frac{1}{3}$

3) $4\frac{2}{5}$

4) $4\frac{2}{8}$

5) $2\frac{1}{5}$

6) $2\frac{8}{11}$

7) $4\frac{4}{7}$

8) $3\frac{7}{12}$

9) $2\frac{1}{3}$

10) $7\frac{5}{7}$

11) $2\frac{7}{10}$

12) $3\frac{4}{9}$

13) $1\frac{5}{8}$

14) $4\frac{3}{11}$

15) $3\frac{4}{7}$

16) $5\frac{2}{8}$

17) $7\frac{1}{7}$

18) $13\frac{1}{2}$

19) $4\frac{2}{7}$

20) $5\frac{2}{10}$

21) $12\frac{1}{3}$

22) $7\frac{1}{8}$

STAAR Subject Test Mathematics Grade 4

Add and Subtract Mixed Numbers

✎ Add mixed numbers.

1) $3\frac{2}{5} + 8\frac{1}{5}$

2) $3\frac{2}{3} + 4\frac{1}{2}$

3) $6\frac{2}{7} + 2\frac{3}{7}$

4) $4\frac{2}{5} + 3\frac{1}{4}$

5) $8\frac{3}{4} - 2\frac{1}{2}$

6) $6\frac{5}{12} - 4\frac{1}{4}$

7) $5\frac{3}{8} - 3\frac{7}{8}$

8) $6\frac{1}{4} - 2\frac{15}{16}$

9) $9\frac{23}{28} - 4\frac{17}{28}$

10) $6\frac{1}{6} + 6\frac{2}{3}$

11) $4\frac{2}{9} + 5\frac{5}{9}$

12) $2\frac{1}{4} + 7\frac{4}{7}$

13) $7\frac{1}{5} - 3\frac{3}{5}$

14) $3\frac{1}{6} + 2\frac{3}{7}$

15) $2\frac{1}{3} + 4\frac{1}{4}$

16) $4\frac{1}{4} - 1\frac{2}{5}$

17) $\frac{1}{3} + 6\frac{1}{6}$

18) $2\frac{3}{5} + 2\frac{1}{10}$

WWW.MathNotion.Com

STAAR Subject Test Mathematics Grade 4

Multiplying Mixed Number

Multiply. Reduce to lowest terms.

1) $2\frac{3}{5} \times 1\frac{3}{4} =$

2) $1\frac{5}{6} \times 1\frac{1}{3} =$

3) $2\frac{3}{5} \times 1\frac{1}{7} =$

4) $3\frac{1}{7} \times 2\frac{1}{2} =$

5) $4\frac{3}{4} \times 1\frac{1}{4} =$

6) $3\frac{1}{2} \times 1\frac{4}{5} =$

7) $3\frac{3}{4} \times 1\frac{1}{2} =$

8) $5\frac{2}{3} \times 3\frac{1}{3} =$

9) $3\frac{2}{3} \times 3\frac{1}{2} =$

10) $2\frac{1}{3} \times 3\frac{1}{2} =$

11) $4\frac{3}{4} \times 3\frac{2}{3} =$

12) $3\frac{2}{4} \times 3\frac{1}{6} =$

13) $2\frac{2}{5} \times 1\frac{1}{3} =$

14) $2\frac{1}{3} \times 1\frac{1}{6} =$

15) $2\frac{2}{3} \times 3\frac{1}{2} =$

16) $2\frac{1}{8} \times 2\frac{2}{5} =$

17) $2\frac{1}{4} \times 1\frac{2}{3} =$

18) $2\frac{3}{5} \times 1\frac{1}{4} =$

19) $2\frac{3}{5} \times 1\frac{5}{6} =$

20) $3\frac{3}{5} \times 2\frac{3}{4} =$

21) $3\frac{3}{4} \times 1\frac{1}{3} =$

22) $2\frac{5}{8} \times 3\frac{1}{4} =$

WWW.MathNotion.Com

STAAR Subject Test Mathematics Grade 4

Answers of Worksheets

Fractions to Mixed Numbers

1) $1\frac{4}{5}$
2) $3\frac{2}{3}$
3) $4\frac{7}{8}$
4) $2\frac{5}{11}$
5) $3\frac{1}{2}$
6) $10\frac{3}{4}$
7) $5\frac{4}{9}$
8) $3\frac{3}{4}$
9) $5\frac{2}{7}$
10) $2\frac{5}{7}$
11) $4\frac{5}{9}$
12) $3\frac{9}{12}$
13) $3\frac{2}{5}$
14) $4\frac{5}{6}$
15) $3\frac{1}{4}$
16) $2\frac{1}{7}$
17) $9\frac{2}{7}$
18) $7\frac{3}{8}$
19) $6\frac{1}{4}$
20) $2\frac{1}{8}$

Mixed Numbers to Fractions

1) $\frac{18}{5}$
2) $\frac{4}{3}$
3) $\frac{22}{5}$
4) $\frac{34}{8}$
5) $\frac{11}{5}$
6) $\frac{30}{11}$
7) $\frac{32}{7}$
8) $\frac{43}{12}$
9) $\frac{7}{3}$
10) $\frac{54}{7}$
11) $\frac{27}{10}$
12) $\frac{31}{9}$
13) $\frac{13}{8}$
14) $\frac{47}{11}$
15) $\frac{25}{7}$
16) $\frac{42}{8}$
17) $\frac{50}{7}$
18) $\frac{27}{2}$
19) $\frac{30}{7}$
20) $\frac{52}{10}$
21) $\frac{37}{3}$
22) $\frac{57}{8}$

Add and Subtract Mixed Numbers

1) $11\frac{3}{5}$
2) $8\frac{1}{6}$
3) $8\frac{5}{7}$
4) $7\frac{13}{20}$
5) $6\frac{1}{4}$
6) $2\frac{1}{6}$
7) $1\frac{1}{2}$
8) $3\frac{5}{16}$
9) $5\frac{3}{14}$
10) $12\frac{5}{6}$
11) $9\frac{7}{9}$
12) $9\frac{23}{28}$
13) $3\frac{3}{5}$
14) $5\frac{25}{42}$
15) $6\frac{7}{12}$
16) $2\frac{17}{20}$
17) $6\frac{1}{2}$
18) $4\frac{7}{10}$

Multiplying Mixed Number

1) $4\frac{11}{20}$
2) $2\frac{4}{9}$
3) $2\frac{34}{35}$
4) $7\frac{6}{7}$
5) $5\frac{15}{16}$
6) $6\frac{3}{10}$
7) $5\frac{5}{8}$
8) $18\frac{8}{9}$
9) $12\frac{5}{6}$

WWW.MathNotion.Com

STAAR Subject Test Mathematics Grade 4

10) $8\frac{1}{6}$

11) $17\frac{5}{12}$

12) $11\frac{1}{12}$

13) $3\frac{1}{5}$

14) $2\frac{13}{18}$

15) $9\frac{1}{3}$

16) $5\frac{1}{10}$

17) $3\frac{3}{4}$

18) $3\frac{1}{4}$

19) $4\frac{23}{30}$

20) $9\frac{9}{10}$

21) 5

22) $8\frac{17}{32}$

STAAR Subject Test Mathematics Grade 4

Chapter 5 : Decimals

Topics that you'll learn in this chapter:

- ✓ Graph Decimal
- ✓ Adding and Subtracting Decimals,
- ✓ Multiplying and Dividing Decimals,
- ✓ Round decimals,
- ✓ Comparing Decimals,

STAAR Subject Test Mathematics Grade 4

Graph Decimals

✏️ Write the decimals indicated by the arrows.

1)

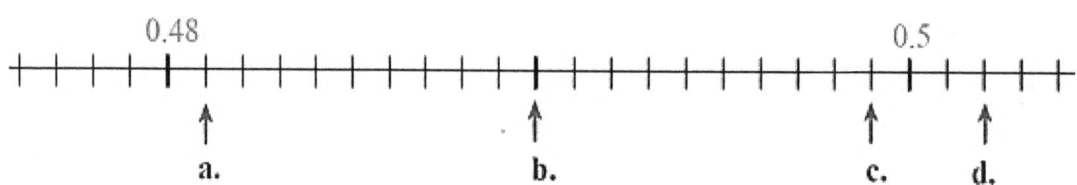

a. _____ b. _____ c. _____ d. _____

2)

0.32   0.33   0.34

a. _____ b. _____ c. _____ d. _____

3)

a. _____ b. _____ c. _____ d. _____

4)

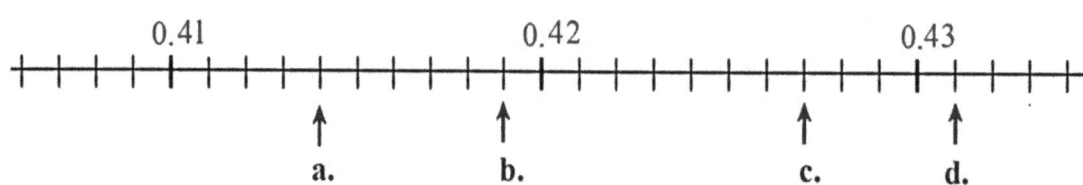

a. _____ b. _____ c. _____ d. _____

WWW.MathNotion.Com

STAAR Subject Test Mathematics Grade 4

Adding and Subtracting Decimals

✎ Add and subtract decimals.

1) 24.19
   -15.42
   $\overline{}$

2) 42.23
   $+25.42$
   $\overline{}$

3) 72.54
   $+11.28$
   $\overline{}$

4) 57.45
   -24.75
   $\overline{}$

5) 43.57
   $+54.85$
   $\overline{}$

6) 86.68
   -54.12
   $\overline{}$

✎ Solve.

7) ____ + 2.7 = 8.1

8) 6.4 + ____ = 12.8

9) 7.9 + ____ = 17

10) 4.6 + ____ = 15.3

11) ____ + 9.4 = 15

12) ____ + 8.24 = 13.54

✎ Order each set of numbers from least to greatest.

1) 0.4, 0.67, 0.44, 0.73, 0.51   ___, ___, ___, ___, ___, ___

2) 3.9, 6.1, 4.28, 7.02, 4.65   ___, ___, ___, ___, ___, ___

3) 1.9, 1.04, 0.79, 0.72, 0.09   ___, ___, ___, ___, ___, ___

4) 2.6, 5.2, 1.9, 4.01, 1.99, 3.2   ___, ___, ___, ___, ___, ___

5) 4.2, 6.1, 3.8, 5.7, 2.1, 2.8   ___, ___, ___, ___, ___, ___

6) 0.56, 0.87, 0.14, 1.24, 3.1   ___, ___, ___, ___, ___, ___

WWW.MathNotion.Com

STAAR Subject Test Mathematics Grade 4

Multiplying and Dividing Decimals

✏️ Find each product.

1) 1.5 × 2.1

4) 6.5 × 0.99

7) 4.8 × 9.1

2) 4.6 × 3.4

5) 12.1 × 5.2

8) 22.35 × 20

3) 6.3 × 2.5

6) 3.4 × 8.9

9) 15.25 × 3.6

✏️ Find each quotient.

10) 3.5 ÷ 0.85

15) 21.5 ÷ 1,000

11) 15.35 ÷ 4.6

16) 4.1 ÷ 100

12) 32.42 ÷ 8.8

17) 9.7 ÷ 10

13) 9.2 ÷ 3.4

18) 6.55 ÷ 1.25

14) 0.84 ÷ 0.1

19) 18.48 ÷ 11.2

WWW.MathNotion.Com

STAAR Subject Test Mathematics Grade 4

Rounding Decimals

✏ Round each decimal number to the nearest place indicated.

1) 0.3̲2

2) 5.0̲1

3) 8.8̲24

4) 0.4̲78

5) 7̲.32

6) 0.2̲9

7) 11.3̲1

8) 6̲.223

9) 9.63̲7

10) 5.4̲804

11) 7̲.9

12) 5̲.2439

13) 6.4̲92

14) 1.6̲2

15) 72̲.85

16) 83̲.67

17) 41.6̲8

18) 794̲.741

19) 52̲.2

20) 76̲.93

21) 3̲.219

22) 72̲.09

23) 486.4̲91

24) 7.0̲8

STAAR Subject Test Mathematics Grade 4

Comparing Decimals

✎ Write the correct comparison symbol (>, < or =).

1) 0.35 ___ 1.8

2) 1.9 ___ 1.19

3) 8.6 ___ 8.6

4) 2.45 ___ 24.5

5) 7.56 ___ 0.756

6) 11.4 ___ 11.05

7) 7.4 ___ 0.7.4

8) 8.56 ___ 0.85

9) 7 ___ 0.7

10) 7.12 ___ 0.712

11) 12.3 ___ 12.5

12) 4.67 ___ 4.68

13) 2.57 ___ 2.75

14) 3.46 ___ 0.346

15) 6.87 ___ 6.78

16) 0.89 ___ 0.98

17) 1.57 ___ 0.157

18) 0.092 ___ 0.091

19) 24.3 ___ 24.3

20) 0.17 ___ 0.71

21) 0.46 ___ 0.64

22) 0.2 ___ 0.08

23) 0.10 ___ 0.1

24) 3.52 ___ 31.5

WWW.MathNotion.Com

STAAR Subject Test Mathematics Grade 4

Answers of Worksheets

Graph Decimals

1) a. 0.481   b. 0.49   c. 0.499   d. 0.502
2) a. 0.321   b. 0.326   c. 0.333   d. 0339
3) a. 0.562   b. 0.569   c. 0.577   d. 0.581
4) a. 0.414   b. 0.419   c. 0.427   d. 0431

Adding and Subtracting Decimals

1) 8.77   4) 32.7   7) 5.4   10) 10.7
2) 67.65   5) 98.42   8) 6.4   11) 5.6
3) 83.82   6) 32.56   9) 9.1   12) 5.3

Order and Comparing Decimals

1) 0.4, 0.44, 0.51, 0.67, 0.73   4) 1.9, 1.99, 2.6, 3.2, 4.01, 5.2
2) 3.9, 4.28, 4.65, 6.1, 7.02   5) 2.1, 2.8, 3.8, 4.2, 5.7, 6.1
3) 0.09, 0.72, 0.79, 1.04, 1.9   6) 0.14, 0.56, 0.87, 1.24, 3.1

Multiplying and Dividing Decimals

1) 3.15   6) 30.26   11) 3.336…   16) 0.041
2) 15.64   7) 43.68   12) 3.684…   17) 0.97
3) 15.75   8) 447   13) 2.705…   18) 5.24
4) 6.435   9) 54.9   14) 8.4   19) 1.65
5) 62.92   10) 4.117… 15) 0.0215

Rounding Decimals

1) 0.3   7) 11.3   13) 6.5   19) 52
2) 5.0   8) 6   14) 1.6   20) 77
3) 8.8   9) 9.64   15) 73   21) 3
4) 0.5   10) 5.5   16) 84   22) 72
5) 7   11) 8   17) 41.7   23) 486.5
6) 0.3   12) 5   18) 795   24) 7.1

Comparing Decimals

1) 0.35 < 1.8   5) 7.56 > 0.756   9) 7 > 0.7
2) 1.9 > 1.19   6) 11.4 > 11.05   10) 7.12 > 0.712
3) 8.6 = 8.6   7) 7.4 > 0.74   11) 12.3 < 12.5
4) 2.45 < 24.5   8) 8.56 > 0.85   12) 4.67 < 4.68

STAAR Subject Test Mathematics Grade 4

13) 2.57 < 2.75
14) 3.46 > 0.346
15) 6.87 > 6.78
16) 0.89 < 0.98
17) 1.57 > 0.157
18) 0.092 > 0.091
19) 24.3 = 24.3
20) 0.17 < 0.71
21) 0.46 < 0.64
22) 0.2 > 0.08
23) 0.10 = 0.1
24) 3.52 < 31.5

STAAR Subject Test Mathematics Grade 4

Chapter 6 : Patterns and Algebraic Thinking

Topics that you'll learn in this chapter:

- ✓ Repeating Patterns,
- ✓ Growing Patterns,
- ✓ Patterns: Numbers,
- ✓ Finding Rules,
- ✓ Algebraic Word Problems,
- ✓ Evaluating Outputs,

Repeating Patterns

✏ Circle the picture that comes next in each picture pattern.

1)

2)

3)

4)

5)

Growing Patterns

✎ Draw the picture that comes next in each growing pattern.

1)

2)

3)

4)

5)

STAAR Subject Test Mathematics Grade 4

Patterns: Numbers

✎ Write the numbers that come next.

1) 2, 5, 8, 11,   ____, ____, ____, ____

2) 10, 15, 20, 25,   ____, ____, ____, ____

3) 4, 8, 12, 16,   ____, ____, ____, ____

4) 7, 17, 27, 37,   ____, ____, ____, ____

5) 5, 12, 19, 26,   ____, ____, ____, ____

6) 8, 16, 24, 32, 40,   ____, ____, ____, ____

✎ Write the next three numbers in each counting sequence.

7) –31, –19, –7, _____, _____, _____, _____

8) 541, 526, 511, _____, _____, _____, _____

9) 14, 34, _____, _____, 94, _____

10) 21, 29, _____, _____, _____

11) 89, 78, _____, _____, _____

12) 95, 82, 69, _____, _____, _____

13) 198, 166, 134, _____, _____, _____

14) What are the next three numbers in this counting sequence?

　　1870, 1970, 2070, _____, _____, _____

15) What is the fourth number in this counting sequence?

　　8, 14, 20, _____

Finding Rules

✎ Complete the output.

1- **Rule:** the output is $x - 10.5$

Input	x	15	18	27	32.25	48.5
Output	y					

1) **Rule:** the output is $x \times 5\frac{1}{3}$

Input	x	3	9	15	21	33
Output	y					

2- **Rule:** the output is $x \div 9$

Input	x	513	387	342	198	126
Output	y					

✎ Find a rule to write an expression.

3- **Rule:** _____

Input	x	4	14	19	24
Output	y	10	35	47.5	60

4- **Rule:** _____

Input	x	5	13	19.6	34.5
Output	y	14.4	22.4	29	43.9

5- **Rule:** _____

Input	x	72	96	132	230.4
Output	y	9	12	16.5	28.8

STAAR Subject Test Mathematics Grade 4

Algebraic Word Problems

Circle the number sentence that fits the problem. Then solve for x.

1) Mary had $42. Then she earned more money (x). Now she has $86.

   $42 + x = $86   OR $42 + $86 = x

   x = ____

2) Lisa had $35. Then she earned more money (x). Now she has $78.

   $35 + x = $78   OR $35 + $78 = x

   x = ____

3) Matthew had $37. Then he earned more money (x). Now he has $98.

   $37 + x = $98   OR $37 + $98 = x

   x = ____

4) Charlotte gave 19 of the cookies he had baked to a friend and now he has 45 cookies left.   45 − 19 = x   OR   x − 19 = 45

   x = ____

5) Mia gave 32 of the cookies she had baked to a friend and now she has 55 cookies left.   55 − 32 = x   OR   x − 32 = 55

   x = ____

6) Lucas gave 41 of the cookies he had baked to a friend and now he has 49 cookies left. .   49 − 41 = x   OR   x − 41 = 49

   x = ____

WWW.MathNotion.Com

STAAR Subject Test Mathematics Grade 4

Evaluating Outputs

✏ Write output of each algebraic expression.

1) $11 - x$, $x = 3$

2) $x + 14$, $x = 4$

3) $8 - 3x$, $x = 1$

4) $3x + \frac{1}{3}$, $x = \frac{1}{2}$

5) $2x + 18$, $x = 1.5$

6) $7 - 3x$, $x = 1.2$

7) $12 + 2x - 15$, $x = 3.5$

8) $25 - 5x$, $x = 2.2$

9) $\frac{44}{x} - 58$, $x = 0.4$

10) $\frac{x}{3} - 15 + x$, $x = 12.6$

11) $\frac{x}{7} + 9.5$, $x = 28.7$

12) $\frac{33}{x} - 5.2 + 2.1x$, $x = 3$

13) $2x - \frac{45}{x} - 12$, $x = 9$

14) $\frac{x}{17} - 1.8$, $x = 34$

15) $2(12.5x + 8)$, $x = 2$

16) $16x + 13x - 27 + 9$,

$x = 1.5$

17) $8.7 - \frac{16}{x} + 3x$,

$x = 4$

18) $5(3a - 2a)$,

$a = 5.5$

19) $14 - 2x + 16 - x$,

$x = 3.5$

20) $6x - 3 - x$,

$x = 1.6$

21) $18 - 2(2x + x)$, $x = 0.2$

WWW.MathNotion.Com

STAAR Subject Test Mathematics Grade 4

Answers of Worksheets

Repeating pattern

1)
2)
3)
4)
5)

Growing patterns

1)
2)
3)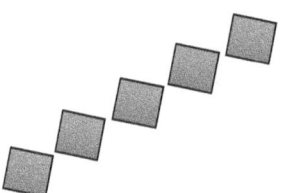

4)
5)

Patterns: Numbers

1) 2, 5, 8, 11, 14, 17, 20, 23
2) 10, 15, 20, 25, 30, 35, 40, 45
3) 4, 8, 12, 16, 20, 24, 28, 32
7) 5, 17, 29, 41
8) 496, 481, 466, 451
9) 14, 34, 54, 74, 94, 114
10) 37, 45, 53
11) 67, 56, 45

4) 7, 17, 27, 37, 47, 57, 67, 77
5) 5, 12, 19, 26, 33, 40, 47, 54
6) 8, 16, 24, 32, 40, 48, 56, 64
12) 56, 43, 30
13) 102, 70, 38
14) 2170, 2270, 2370
15) 26

Finding Rules

1)

Input	x	15	18	27	32.25	48.5
Output	y	4.5	7.5	16.5	21.75	38

2)

Input	x	3	9	15	21	33
Output	y	16	48	80	112	176

WWW.MathNotion.Com

STAAR Subject Test Mathematics Grade 4

3)
Input	x	513	387	342	198	126
Output	y	57	43	38	22	14

4) $y = 2.5x$ 5) $y = x + 9.4$ 6) $y = x \div 8$

Algebraic Word Problems

1) $\$42 + x = \$86; x = 44$ 4) $x - 19 = 45; x = 64$
2) $\$35 + x = \$78; x = 43$ 5) $x - 32 = 55; x = 87$
3) $\$37 + x = \$98; x = 61$ 6) $x - 41 = 49; x = 90$

Evaluating Outputs

1) 8   7) 4   13) 1   19) 19.5
2) 18   8) 14   14) 0.2   20) 5
3) 5   9) 52   15) 66   21) 16.8
4) $\frac{11}{6}$ 10) 1.8   16) 25.5
5) 21   11) 13.6   17) 16.7
6) 3.4   12) 12.1   18) 27.5

WWW.MathNotion.Com

STAAR Subject Test Mathematics Grade 4

STAAR Subject Test Mathematics Grade 4

Chapter 7 : Measurement

Topics that you'll learn in this chapter:

- ✓ Reference Measurement Units,
- ✓ Metric Length Units,
- ✓ Customary Length Units,
- ✓ Metric Capacity Units,
- ✓ Customary Capacity Units,
- ✓ Metric Weight and Mass Units,
- ✓ Customary Weight and Mass Units,
- ✓ Temperature Units,
- ✓ Time,
- ✓ Add Money Amounts,
- ✓ Subtract Money Amounts,
- ✓ Money: Word Problems,

STAAR Subject Test Mathematics Grade 4

Reference Measurement Units

LENGTH

Customary

1 mile (mi) = 1,760 yards (yd)

1 yard (yd) = 3 feet (ft)

1 foot (ft) = 12 inches (in.)

Metric

1 kilometer (km) = 1,000 meters (m)

1 meter (m) = 100 centimeters (cm)

1 centimeter (cm) = 10 millimeters (mm)

VOLUME AND CAPACITY

Customary

1 gallon (gal) = 4 quarts (qt)

1 quart (qt) = 2 pints (pt.)

1 pint (pt.) = 2 cups (c)

1 cup (c) = 8 fluid ounces (Fl oz)

Metric

1 liter (L) = 1,000 milliliters (mL)

WEIGHT AND MASS

Customary

1 ton (T) = 2,000 pounds (lb.)

1 pound (lb.) = 16 ounces (oz)

Metric

1 kilogram (kg) = 1,000 grams (g)

1 gram (g) = 1,000 milligrams (mg)

Time

1 year = 12 months

1 year = 52 weeks

1 week = 7 days

1 day = 24 hours

1 hour = 60 minutes

1 minute = 60 seconds

STAAR Subject Test Mathematics Grade 4

Metric Length Units

✏️ Convert to the units.

1) 300 mm = _____ cm

2) 8 m = _____ mm

3) 4.5 m = _____ cm

4) 7 km = _____ m

5) 9,400 mm = _____ m

6) 1,100 cm = _____ m

7) 2.8 m = _____ cm

8) 4,000 mm = _____ cm

9) 7,000 mm = _____ m

10) 2 km = _____ mm

11) 14.9 km = _____ m

12) 20 m = _____ cm

13) 5,000 m = _____ km

14) 7,600 m = _____ km

Customary Length Units

✏️ Convert to the units.

1) 8 ft = _____ in

2) 4 ft = _____ in

3) 6 yd = _____ ft

4) 10 yd = _____ ft

5) 3,520 yd = _____ mi

6) 60 in = _____ ft

7) 144 in = ____ yd

8) 0.5 mi = _____ yd

9) 15 yd = _____ in

10) 42 yd = _____ in

11) 99 ft = _____ yd

12) 1.5 mi = _____ yd

13) 84 in = _____ ft

14) 30 yd = _____ feet

STAAR Subject Test Mathematics Grade 4

Metric Capacity Units

✎ Convert the following measurements.

1) 32.4 l = _____ ml

2) 7.1 l = _____ ml

3) 54 l = _____ ml

4) 92 l = _____ ml

5) 48 l = _____ ml

6) 13 l = _____ ml

7) 750 ml = _____ l

8) 2,400 ml = _____ l

9) 73,000 ml = _____ l

10) 8,000 ml = _____ l

11) 49,000 ml = _____ l

12) 5,500 ml = _____ l

Customary Capacity Units

✎ Convert the following measurements.

1) 51 gal = _____ qt.

2) 35 gal = _____ pt.

3) 68 gal = _____ c.

4) 20 pt. = _____ c

5) 12.5 qt = _____ pt.

6) 22.5 qt = _____ c

7) 51 pt. = _____ c

8) 48 c = _____ gal

9) 96 pt. = _____ gal

10) 136 qt = _____ gal

11) 15 c = _____ fl oz

12) 44 c = _____ qt

13) 240 c = _____ pt.

14) 148 qt = _____ gal

15) 160 pt. = _____ qt

16) 104 fl oz = _____ c.

WWW.MathNotion.Com

STAAR Subject Test Mathematics Grade 4

Metric Weight and Mass Units

Convert.

1) 60 kg = _____ g

2) 24 kg = _____ g

3) 610 kg = _____ g

4) 82 kg = _____ g

5) 95.8 kg = _____ g

6) 4.85 kg = _____ g

7) 1.5 kg = _____ g

8) 95,000 g = _____ kg

9) 241,000 g = _____ kg

10) 700,000 g = _____ kg

11) 2,500 g = _____ kg

12) 28,900 g = _____ kg

13) 970,000 g = _____ kg

14) 325,500 g = _____ kg

Customary Weight and Mass Units

Convert.

1) 10,000 lb. = _____ T

2) 19,000 lb. = _____ T

3) 32,000 lb. = _____ T

4) 16,800 lb. = _____ T

5) 27 lb. = _____ oz

6) 25.4 lb. = _____ oz

7) 124 lb. = _____ oz

8) 4T = _____ lb.

9) 7T = _____ lb.

10) 11.2T = _____ lb.

11) 12.8T = _____ lb.

12) $\frac{6}{5}$ T = _____ oz

13) 9.125 T = _____ oz

14) $\frac{3}{4}$ T = _____ oz

WWW.MathNotion.Com

STAAR Subject Test Mathematics Grade 4

Time

✏ Convert to the units.

1) 22 hr. = _____ min

2) 12.5 year = _____ week

3) 5.2hr = _____ sec

4) 21min = _____ sec

5) 1,800min = _____ hr.

6) 730 day = _____ year

7) 1.5year = _____ hr.

8) 42 day = _____ hr.

9) 2 day = _____ min

10) 660 min = _____ hr.

11) 10year = _____ month

12) 2,124sec = _____ min

13) 168 hr = _____ day

14) 19 weeks = _____ day

✏ How much time has passed?

1) From 2:25 A.M. to 5:35 A.M.: ____ hours and ___ minutes.

2) From 3:30 A.M. to 7:55 A.M.: ____ hours and ___ minutes.

3) It's 6:30 P.M. What time was 2 hours ago? _____ O'clock

4) 4:30 A.M to 7:50 AM: _____ hours and _____ minutes.

5) 4:15 A.M to 8:35 AM: _____ hours and _____ minutes.

6) 5:10 A.M. to 7:35 AM. = _____ hour(s) and _____ minutes.

7) 10:55 A.M. to 3:25 PM. = _____ hour(s) and _____ minutes

8) 8:05 A.M. to 8:40 A.M. = _____ minutes

9) 6:02 A.M. to 6:49 A.M. = _____ minutes

Money Amounts

✎ Add.

1) $128 + $328
   $225 + $245
   $150 + $186

2) $453 + $128
   $440 + $541
   $258 + $248

3) $645 + $112.5
   $235.4 + $452.1
   $125.99 + $148.32

4) $321.40 + $175.80
   $458.10 + $752.65
   $652.00 + $324.70

✎ Subtract.

5) $725 − $334
   $543 − $248
   $349 − $122

6) $658.20 − $220.30
   $752.10 − $452.15
   $312.50 − $89.90

7) $315.90 − $220.10
   $548.40 − $342.10
   $968.40 − $324.50

8) Linda had $18.60. She bought some game tickets for $9.25. How much did she have left?

> STAAR Subject Test Mathematics Grade 4

Money: Word Problems

🖎 Solve.

1) How many boxes of envelopes can you buy with $40 if one box costs $8?

2) After paying $7.22 for a salad, Ella has $45.86. How much money did she have before buying the salad?

3) How many packages of diapers can you buy with $96 if one package costs $6?

4) Last week James ran 28.5 miles more than Michael. James ran 59 miles. How many miles did Michael run?

5) Last Friday Jacob had $14.68. Over the weekend he received some money for cleaning the attic. He now has $38.95. How much money did he receive?

6) After paying $3.15 for a sandwich, Amelia has $48.69. How much money did she have before buying the sandwich?

STAAR Subject Test Mathematics Grade 4

Answers of Worksheets

Metric length

1) 30 cm
2) 8,000 mm
3) 450 cm
4) 7,000 m
5) 9.4 m
6) 11 m
7) 280 cm
8) 400 cm
9) 7 m
10) 2,000,000 mm
11) 14,900 m
12) 2,000 cm
13) 5 km
14) 7.6 km

Customary Length

1) 96
2) 48
3) 18
4) 30
5) 2
6) 5
7) 4
8) 880
9) 540
10) 1,512
11) 33
12) 2,640
13) 7
14) 90

Metric Capacity

1) 32,400 ml
2) 7,100 ml
3) 54,000 ml
4) 92,000 ml
5) 48,000 ml
6) 13,000 ml
7) 0.75ml
8) 2.4 ml
9) 73 ml
10) 8L
11) 49 L
12) 5.5 L

Customary Capacity

1) 204 qt
2) 280 pt.
3) 1,088 c
4) 40 c
5) 25 pt.
6) 90c
7) 102 c
8) 3 gal
9) 12 gal
10) 34 gal
11) 120 qt
12) 11qt
13) 120 pt.
14) 37 gal
15) 80 qt
16) 13 pt.

Metric Weight and Mass

1) 60,000 g
2) 24,000 g
3) 610,000 g
4) 82,000 g
5) 95,800g
6) 4,850 g
7) 1,500 g
8) 95 kg
9) 241 kg
10) 700 kg
11) 2.5 kg
12) 28.9 kg
13) 970 kg
14) 325.5 kg

Customary Weight and Mass

1) 5 T
2) 9.5 T
3) 16 T
4) 8.4 T
5) 432 oz
6) 406.4 oz

www.MathNotion.Com

STAAR Subject Test Mathematics Grade 4

7) 1,984 oz
8) 8,000 lb.
9) 14,000 lb.

10) 22,400 lb.
11) 25,600 lb.
12) 38,400 oz

13) 292,000 oz
14) 24,000 oz

Time - Convert

1) 1,320 min
2) 650 weeks
3) 18,720 sec
4) 1,260 sec
5) 30 hr

6) 2 year
7) 13,140 hr
8) 1,008 hr
9) 2,880 min
10) 11hr

11) 120 months
12) 35.4 min
13) 7 days
14) 133 days

Time - Gap

1) 3:10
2) 4:25
3) 4:30P.M.

4) 3:20
5) 4:20
6) 2:25

7) 4:30
8) 35 minutes
9) 47 minutes

Add Money

1) 456, 470, 336
2) 581, 981, 506

3) 757.5, 687.5, 274.31
4) 497.2, 1210.75, 976.7

Subtract Money

5) 391, 295, 227
6) 437.9, 299.95, 222.6

7) 95.8, 206.3, 643.9
8) $9.35

Money: word problem

1) 5
2) $53.08

3) 16
4) 30.5

5) 24.27
6) 51.84

STAAR Subject Test Mathematics Grade 4

Chapter 8 : Geometric

Topics that you'll learn in this chapter:

- ✓ Identifying Angles,
- ✓ Estimate and Measure Angles,
- ✓ Polygon Names,
- ✓ Classify Triangles,
- ✓ Parallel Sides in Quadrilaterals,
- ✓ Identify Parallelograms,
- ✓ Identify Trapezoids,
- ✓ Identify Rectangles,
- ✓ Perimeter and Area of Squares,
- ✓ Perimeter and Area of rectangles,
- ✓ Area and Perimeter: Word Problems,
- ✓ Volume,

Identifying Angles

✎ Write the name of the angles (Acute, Right, Obtuse, and Straight).

1)

2)

3)

4)

5)

6)

7)

8)

STAAR Subject Test Mathematics Grade 4

Estimate Angle Measurements

🖎 Estimate the approximate measurement of each angle in degrees.

1)

2)

3)

4)

5)

6)

7)

8)

WWW.MathNotion.Com

STAAR Subject Test Mathematics Grade 4

Measure Angles with a Protractor

✎ Use protractor to measure the angles below.

1)

2)

3)

4)

✎ Use a protractor to draw angles for each measurement given.

1) 140°

2) 100°

3) 110°

4) 120°

5) 55°

Polygon Names

✎ Write name of polygons.

1)

2)

3)

4)

5)

6)

Classify Triangles

✎ Classify the triangles by their sides and angles.

1)

2)

3)

4)

5)

6)

Parallel Sides in Quadrilaterals

✏ Write name of quadrilaterals.

1)

2)

3)

4)

5)

6)

STAAR Subject Test Mathematics Grade 4

Identify Rectangles

✏️ Solve.

1) A rectangle has _____ sides and _____ angles.

2) Draw a rectangle that is 5.5 centimeters long and 2.5 centimeters wide. What is the perimeter?

3) Draw a rectangle 3.5 cm long and 1.5 cm wide.

4) Draw a rectangle whose length is 4.25 cm and whose width is 2.45 cm. What is the perimeter of the rectangle?

5) What is the perimeter of the rectangle?

7.2

5.8

STAAR Subject Test Mathematics Grade 4

Perimeter: Find the Missing Side Lengths

✏ Find the missing side of each shape.

1) perimeter = 57.2

2) perimeter = 21.2

3) perimeter = 27.5

4) perimeter = 35.2

5) perimeter = 75.6

6) perimeter = 30.8

7) perimeter = 36.25

8) perimeter = 46.8

STAAR Subject Test Mathematics Grade 4

Perimeter and Area of Squares

✍ Find perimeter and area of squares.

1) A: _____ , P: _____

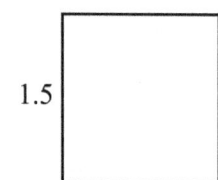

2) A: _____ , P: _____

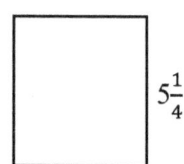

3) A: _____ , P: _____

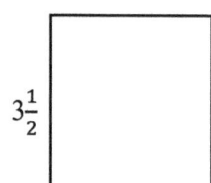

4) A: _____ , P: _____

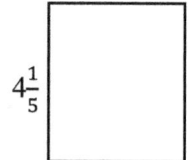

5) A: _____ , P: _____

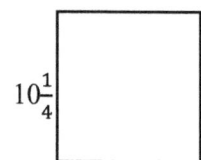

6) A: _____ , P: _____

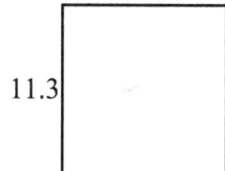

7) A: _____ , P: _____

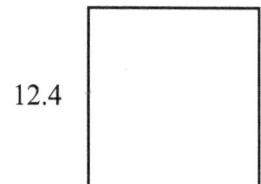

8) A: _____ , P: _____

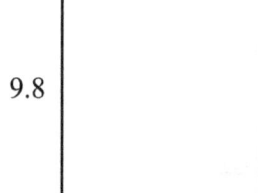

STAAR Subject Test Mathematics Grade 4

Perimeter and Area of rectangles

✎ Find perimeter and area of rectangles.

1) A: _____, P: _____

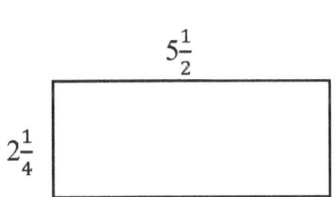
$5\frac{1}{2}$ / $2\frac{1}{4}$

2) A: _____, P: _____

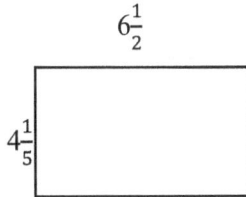
$6\frac{1}{2}$ / $4\frac{1}{5}$

3) A: _____, P: _____

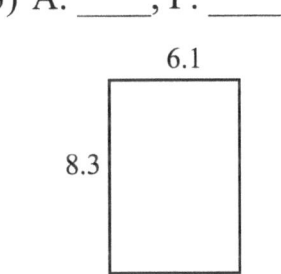
6.1 / 8.3

4) A: _____, P: _____

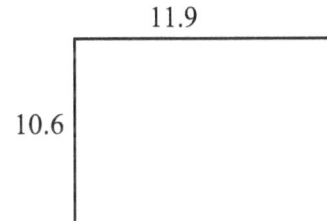
11.9 / 10.6

5) A: _____, P: _____

7.6 / $5\frac{1}{4}$

6) A: _____, P: _____

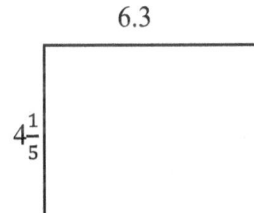
6.3 / $4\frac{1}{5}$

7) A: _____, P: _____

9.7 / $3\frac{1}{2}$

8) A: _____, P: _____

11.4 / 5.8

WWW.MathNotion.Com

STAAR Subject Test Mathematics Grade 4

Find the Area or Missing Side Length of a Rectangle

✎ Find area or missing side length of rectangles.

1) Area =?

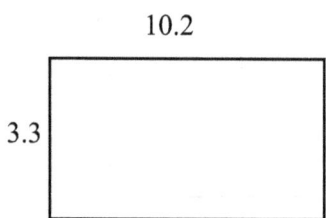

2) Area = 42.12, x=?

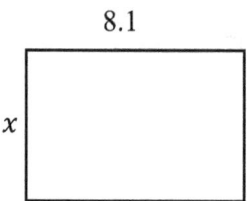

3) Area = 29.52, x=?

4) Area =?

5) Area =?

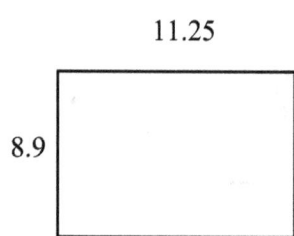

6) Area = 662.34 x=?

7) Area = 216.24, x=?

8) Area 336.42, x=?

Area and Perimeter: Word Problems

Solve.

1) The area of a rectangle is 90.86 square meters. The width is 7.7 meters. What is the length of the rectangle?

2) A square has an area of 6.25 square feet. What is the perimeter of the square?

3) Ava built a rectangular vegetable garden that is 3.2 feet long and has an area of 21.12 square feet. What is the perimeter of Ava's vegetable garden?

4) A square has a perimeter of 12.8 millimeters. What is the area of the square?

5) The perimeter of David's square backyard is 0.96 meters. What is the area of David's backyard?

6) The area of a rectangle is 37.63 square inches. The length is 7.1 inches. What is the perimeter of the rectangle?

STAAR Subject Test Mathematics Grade 4

Volume of Cubes and Rectangle Prisms

✏️ **Find the volume of each of the rectangular prisms.**

1)

2)

3)

4)

5)

6)
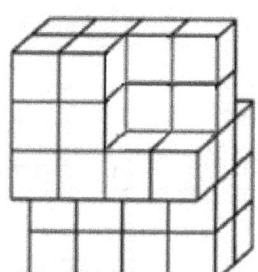

STAAR Subject Test Mathematics Grade 4

Answers of Worksheets

Identifying Angles

1) Right 3) Obtuse 5) Acute 7) Obtuse
2) Acute 4) Straight 6) Obtuse 8) Acute

Estimate Angle Measurements

1) 160° 3) 110° 5) 130° 7) 90°
2) 180° 4) 75° 6) 45° 8) 60°

Measure Angles with a Protractor

1) 50° 2) 135° 3) 20° 4) 170°

Draw Angles

1) 2) 3)

4) 5)

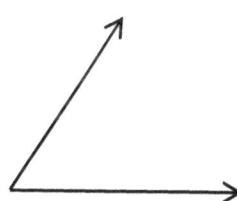

Polygon Names

1) Diamond 3) Pentagon 5) Heptagon
2) Parallelogram 4) Trapezius 6) Octagon

Classify Triangles

1) Scalene, acute 4) Scalene, right
2) Isosceles, acute 5) Isosceles, right
3) Equilateral, acute 6) Scalene, obtuse

STAAR Subject Test Mathematics Grade 4

Parallel Sides in Quadrilaterals

1) Hexagon
2) Kike
3) Parallelogram
4) Trapezoid
5) Rhombus
6) Rectangle

Identify Rectangles

1) 4 - 4
2) 16
3) Draw the rectangle.
4) 13.4
5) 26

Perimeter: Find the Missing Side Lengths

1) 14.3
2) 7.2
3) 5.5
4) 8.8
5) 18.9
6) 6.2
7) 7
8) 11.7

Perimeter and Area of Squares

1) A: 2.25, P: 6
2) A: 27.56, P: 21
3) A: 12.25, P: 14
4) A: 17.64, P: 16.8
5) A: 105.063 P: 41
6) A: 127.69, P: 45.2
7) A: 153.76, P: 49.6
8) A: 96.04, P: 39.2

Perimeter and Area of rectangles

1) A: 12.375, P: 15.5
2) A: 27.3, P: 21.4
3) A: 50.63, P: 28.8
4) A: 126.14, P: 45
5) A: 39.9, P: 25.7
6) A: 26.46, P: 21
7) A: 33.95, P: 26.4
8) A: 66.12, P: 34.4

Find the Area or Missing Side Length of a Rectangle

1) 33.66
2) 5.2
3) 7.2
4) 60.125
5) 100.125
6) 28.5
7) 10.2
8) 17.8

Area and Perimeter: Word Problems

1) 11.8
2) 10
3) 19.6
4) 10.24
5) 0.0576
6) 24.8

Volume of Cubes and Rectangle Prisms

1) 452.025 cm^3
2) 429.2 cm^3
3) 148.877 c m^3
4) 583.848 cm^3
5) 32
6) 40

Chapter 9 : Data Graphs, and Statistics

Topics that you'll learn in this chapter:

- ✓ Bar Graph,
- ✓ Tally and Pictographs,
- ✓ Dot Plots
- ✓ Line Graphs,
- ✓ Stem-And-Leaf Plot,
- ✓ Coordinate Plane,

Bar Graph

🖊 Graph the given information as a bar graph.

Day	Hot dogs sold
Monday	50
Tuesday	80
Wednesday	10
Thursday	30
Friday	70

Tally and Pictographs

✎ Using the key, draw the pictograph to show the information.

Key: = 2 animals

STAAR Subject Test Mathematics Grade 4

Dot plots

The ages of students in a Math class are given below.

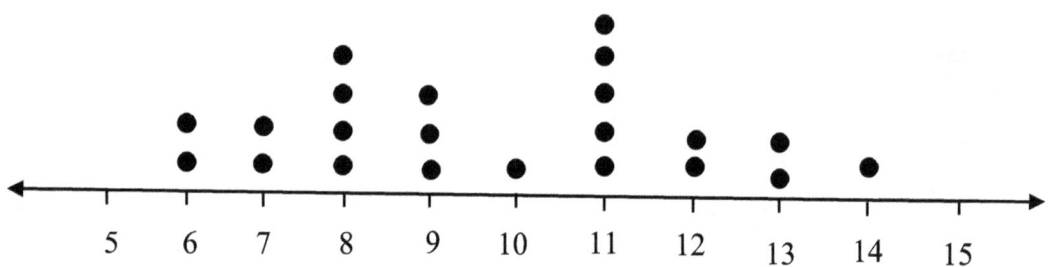

1) What is the total number of students in math class?

2) How many students are at least 12 years old?

3) Which age(s) has the most students?

4) Which age(s) has the fewest student?

5) Determine the median of the data.

6) Determine the range of the data.

7) Determine the mode of the data.

Line Graphs

David work as a salesman in a store. He records the number of shoes sold in five days on a line graph. Use the graph to answer the question.

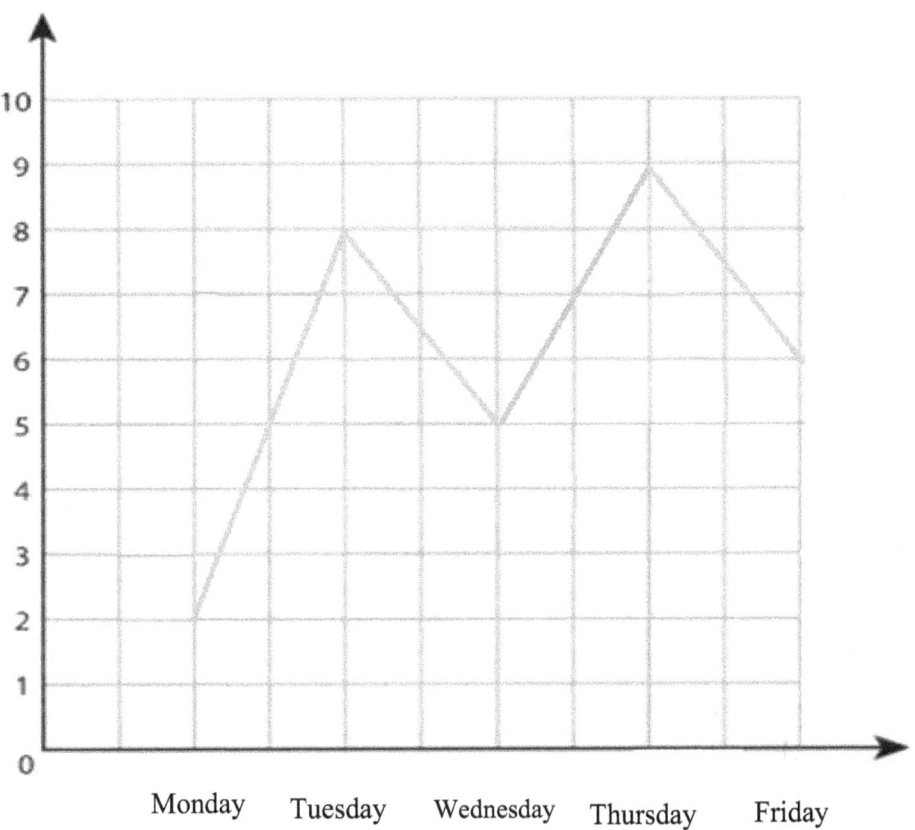

1) How many shoes were sold on Friday?

2) Which day had the minimum sales of shoes?

3) Which day had the maximum number of shoes sold?

4) How many shoes were sold in 5 days?

STAAR Subject Test Mathematics Grade 4

Stem–And–Leaf Plot

✏ Make stem ad leaf plots for the given data.

1) 42, 47, 14, 19, 42, 69, 65, 49, 42, 10, 64

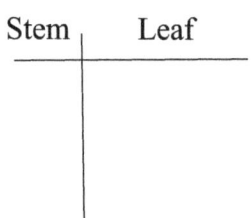

2) 43, 85, 52, 48, 45, 43, 51, 81, 59, 50, 85, 89

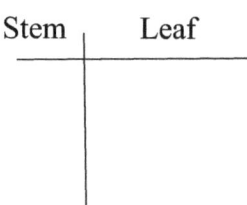

3) 112, 39, 46, 35, 80, 119, 42, 114, 37, 112, 47, 119

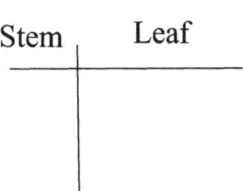

4) 90, 50, 131, 93, 112, 56, 139, 98, 115, 59, 98, 135, 111

Stem | Leaf

WWW.MathNotion.Com

Coordinate Plane

🖎 Plot each point on the coordinate grid.

1) A (4, 6)   3) C (1, 5)   5) E (4, 8)
2) B (3, 2)   4) D (5, 7)   6) F (9, 2)

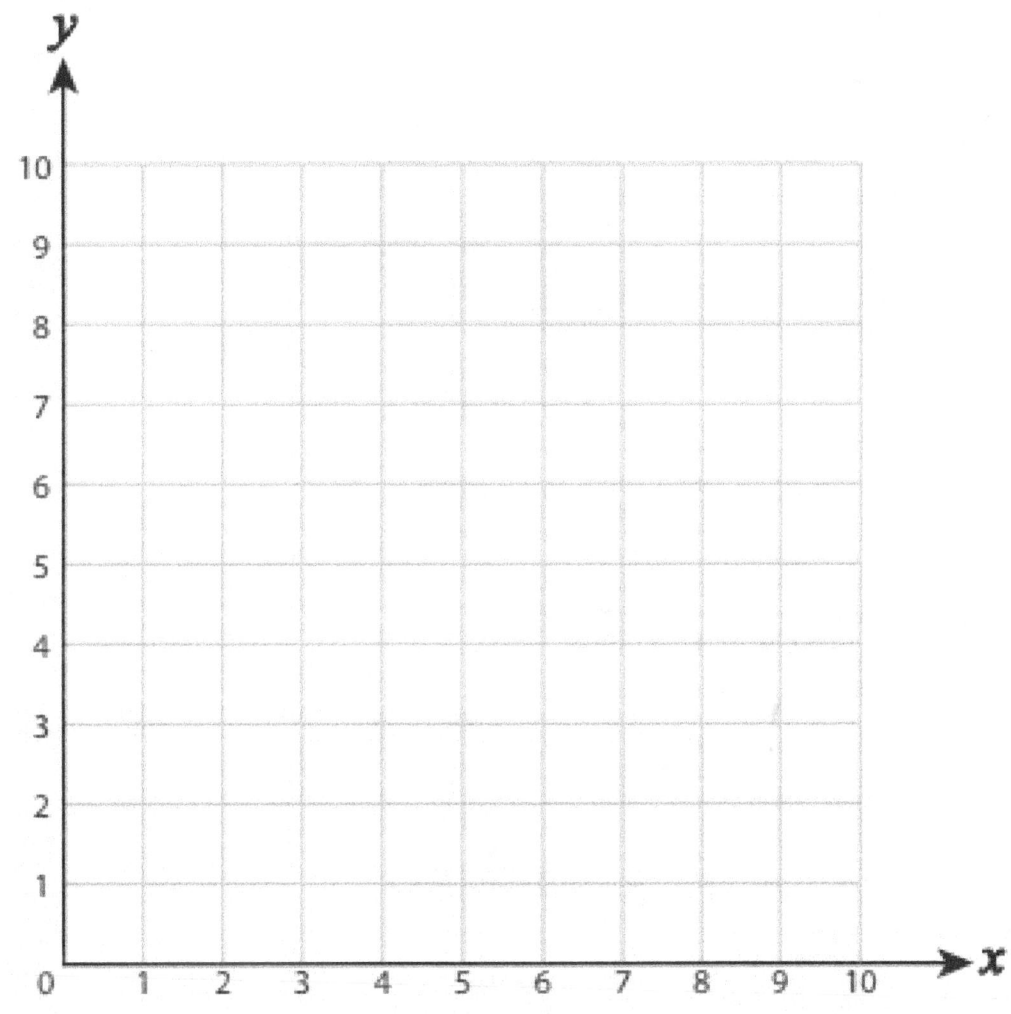

STAAR Subject Test Mathematics Grade 4

Answers of Worksheets

Bar Graph

Tally and Pictographs

Dot plots

1) 22
2) 5
3) 11
4) 10 and 14
5) 2
6) 4
7) 2

STAAR Subject Test Mathematics Grade 4

Line Graphs

1) 6   2) Monday   3) Thursday   4) 30

Stem–And–Leaf Plot

1)
Stem	leaf
1	0 4 9
4	2 2 2 7 9
6	4 5 9

2)
Stem	leaf
4	3 3 5 8
5	0 1 2 9
8	1 5 5 9

3)
Stem	leaf
3	5 7 9
4	2 6 7
8	0
11	2 2 4 9 9

4)
Stem	leaf
5	0 6 9
9	0 3 8 8
11	1 2 5
13	1 5 9

Coordinate Plane

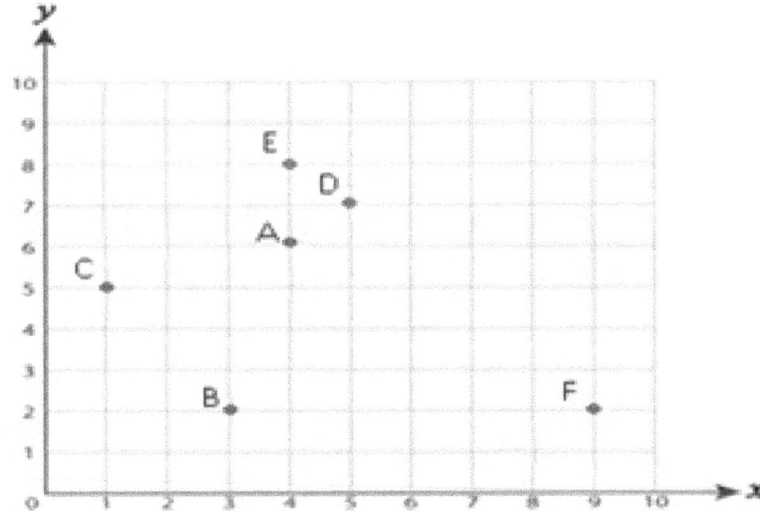

STAAR Subject Test Mathematics Grade 4

Chapter 10 : Three-Dimensional Figures

Topics that you'll learn in this chapter:

- ✓ Identify Three–Dimensional Figures,
- ✓ Count Vertices, Edges, and Faces,
- ✓ Identify Faces of Three–Dimensional Figures,

Identify Three–Dimensional Figures

✍ Write the name of each shape.

1)

2)

3)

4)

5)

6)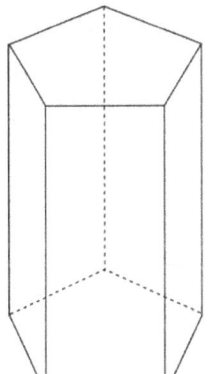

Count Vertices, Edges, and Faces

	Shape	Number of edges	Number of faces	Number of vertices
1)		___	___	___
2)		___	___	___
3)		___	___	___
4)		___	___	___
5)		___	___	___
6)		___	___	___

Identify Faces of Three–Dimensional Figures

✏ Write the number of faces.

1)

2)

3)

4)

5)

6)

7)

8)

STAAR Subject Test Mathematics Grade 4

Answers of Worksheets

Identify Three–Dimensional Figures

1) Cube
2) Triangular pyramid
3) Triangular prism
4) Square pyramid
5) Rectangular prism
6) Pentagonal prism
7) Hexagonal prism

Count Vertices, Edges, and Faces

Shape	Number of edges	Number of faces	Number of vertices
1)	6	4	4
2)	8	5	5
3)	12	6	8
4)	15	7	10
5)	12	6	8
6)	18	8	12

Identify Faces of Three–Dimensional Figures

1) 6
2) 2
3) 5
4) 4
5) 6
6) 7
7) 8
8) 5

Chapter 11 : Symmetry and Transformations

Topics that you'll learn in this chapter:

- ✓ Line Segments,
- ✓ Identify Lines of Symmetry,
- ✓ Count Lines of Symmetry,
- ✓ Parallel, Perpendicular and Intersecting Lines,

STAAR Subject Test Mathematics Grade 4

Line Segments

✏ Write each as a line, ray, or line segment.

1)

2)

3)

4)

5)

6)

7)

8)

WWW.MathNotion.Com

Identify Lines of Symmetry

🖉 Tell whether the line on each shape a line of symmetry is.

1)

2)

3)

4)

5)

6)

7)

8)

Count Lines of Symmetry

✏️ Draw lines of symmetry on each shape. Count and write the lines of symmetry you see.

1)

2)

3)

4)

5)

6)

7)

8)

STAAR Subject Test Mathematics Grade 4

Parallel, Perpendicular and Intersecting Lines

✍ State whether the given pair of lines are parallel, perpendicular, or intersecting.

1)

2)

3)

4)

5)

6)

7)

8)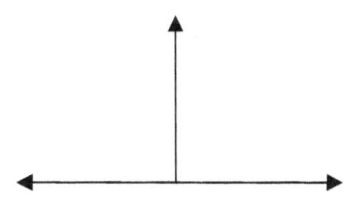

STAAR Subject Test Mathematics Grade 4

Answers of Worksheets

Line Segments

1) Ray   4) Ray   7) Line
2) Line segment 5) Ray   8) Line segment
3) Line   6) Line

Identify lines of symmetry.

1) yes   3) no   5) yes   7) no
2) no   4) yes   6) yes   8) yes

Count lines of symmetry.

1)   2)   3)   4)

5)   6)   7)   8)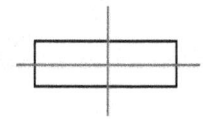

Parallel, Perpendicular and Intersecting Lines

1) Intersection   5) Intersection
2) Perpendicular   6) Perpendicular
3) Parallel   7) Parallel
4) Intersection   8) Perpendicular

STAAR Subject Test Mathematics Grade 4

Chapter 12 : STAAR Math Practice Tests

Time to Test

Time to refine your skill with a practice examination.

Take a REAL STAAR Mathematics test to simulate the test day experience. After you've finished, score your test using the answer key.

Before You Start

- You'll need a pencil and scratch papers to take the test.
- For this practice test, don't time yourself. Spend time as much as you need.
- It's okay to guess. You won't lose any points if you're wrong.
- After you've finished the test, review the answer key to see where you went wrong.

Calculators are not permitted for Grade 4 STAAR Tests

Good Luck!

STAAR Subject Test Mathematics Grade 4

STAAR Subject Test Mathematics Grade 4

STAAR GRADE 4 MAHEMATICS REFRENCE MATERIALS

LENGTH

Customary	Metric
1 mile (mi) = 1,760 yards (yd)	1 kilometer (km) = 1,000 meters (m)
1 yard (yd) = 3 feet (ft)	1 meter (m) = 100 centimeters (cm)
1 foot (ft) = 12 inches (in.)	1 centimeter (cm) = 10 millimeters (mm)

VOLUME AND CAPACITY

Customary	Metric
1 gallon (gal) = 4 quarts (qt)	1 liter (L) = 1,000 milliliters (mL)
1 quart (qt) = 2 pints (pt.)	
1 pint (pt.) = 2 cups (c)	
1 cup (c) = 8 fluid ounces (Fl oz)	

WEIGHT AND MASS

Customary	Metric
1 ton (T) = 2,000 pounds (lb.)	1 kilogram (kg) = 1,000 grams (g)
1 pound (lb.) = 16 ounces (oz)	1 gram (g) = 1,000 milligrams (mg)

Time

1 year = 12 months	1 day = 24 hours
1 year = 52 weeks	1 hour = 60 minutes
1 week = 7 days	1 minute = 60 seconds

Perimeter

Square		$P = 4S$
Rectangle	$P = L + W + L + W$ or	$P = 2L + 2W$

Area

Square	$A = S \times S$
Rectangle	$A = L \times W$

STAAR Subject Test Mathematics Grade 4

State of Texas Assessments of Academic Readiness

STAAR Practice Test 1

Mathematics

GRADE 4

Released Month Year

STAAR Subject Test Mathematics Grade 4

1) Which number correctly completes the number sentence 50 × 32 =?

   A. 3,600

   B. 800

   C. 1,800

   D. 1,600

2) Tam has 350 cards. He wants to put them in boxes of 70 cards. How many boxes does he need?

   A. 3

   B. 5

   C. 8

   D. 6

3) If this clock shows a time in the noon, what time was it 2 hours and 45 minutes ago?

   A. 3:45 AM

   B. 3:30 AM

   C. 2:30 AM

   D. 2:45 AM

4) Joe has 679 crayons. What is this number rounded to the nearest ten?

   (Write your answer in the box below).

5) Use a rule to measure the length and width of the following rectangle to the nearest inches.

What measurement is the closest to the area of the rectangle in square inches?

A. 12 square inches

B. 15 square inches

C. 28 square inches

D. 10 square inches

6) There are 7 days in a week. There are 31 days in the month of July. How many times as many days are there in July than are in one week?

A. 5 times

B. 6 times

C. 4 times

D. 3 times

STAAR Subject Test Mathematics Grade 4

7) A football team is buying new uniforms. Each uniform cost $19. The team wants to buy 15 uniforms.

Which equation represents a way to find the total cost of the uniforms?

A. $(19 \times 15) + (10 \times 5) = 285 + 50$

B. $(19 \times 6) + (10 \times 15) = 109 + 150$

C. $(19 \times 10) + (19 \times 5) = 190 + 95$

D. $(19 \times 10) + (5 \times 15) = 190 + 75$

8) A number sentence such as $20 + Z = 55$ can be called an equation. If this equation is true, then which of the following equations is not true?

A. $55 - 20 = Z$

B. $55 - Z = 20$

C. $Z - 20 = 55$

D. $Z + 20 = 55$

9) Ella described a number using these clues:

Three – digit even numbers that have a 5 in the hundreds place and a 3 in the tens place. Which number could fit Ella's description?

A. 533

B. 532

C. 535

D. 537

10) Use the table below to answer the question.

Favorite Sports

Sport	Number of Votes
football (FB)	9
basketball (BB)	4
soccer (SOC)	10
volleyball (VB)	2

The students in the fourth-grade class voted for their favorite sport. Which bar graph shows results of the students vote?

A.

B.

C.

D.
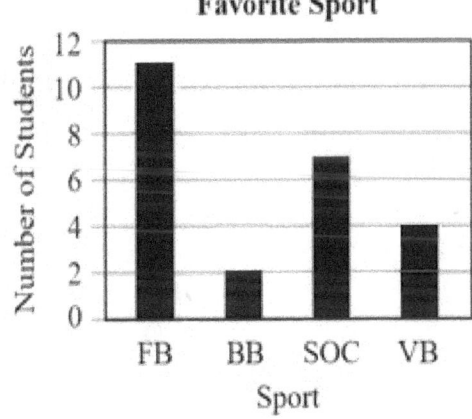

STAAR Subject Test Mathematics Grade 4

11) Use the picture below to answer the question.

Which decimal number names the shaded part of this square?

A. 0.32

B. 0.68

C. 0.34

D. 0.66

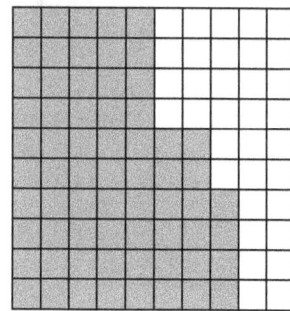

12) Use the table below to answer the question.

Which list of city populations is in order from least to greatest?

A. 30,032; 21,851; 37,009; 28,081.

B. 21,851; 28,081; 30,032; 37,009.

C. 28,081; 30,032; 37,009; 21,851.

D. 37,009; 30,032; 28,081; 21,851.

City Populations	
City	Population
Denton	21,851
Bomberg	30,032
Windham	28,081
Sanhill	37,009

13) Which number correctly completes the subtraction sentence 6.0 – 4.75 = _____ ?

A. 1.15

B. 1.25

C. 1.50

D. 2.50

STAAR Subject Test Mathematics Grade 4

14) For a concert, there are children's tickets and adult tickets for sale. Of the total available tickets, $\frac{52}{100}$ have been sold as adult tickets and $\frac{1}{10}$ as children's tickets.

The rest of the tickets have not been sold.

What fraction of the total number of tickets for the concert have been sold?

A. $1\frac{1}{52}$

B. $\frac{31}{50}$

C. $\frac{31}{52}$

D. $\frac{52}{10}$

15) Circle a reasonable measurement for the angle F:

A. 125°

B. 60°

C. 45°

D. 190°

16) Peter's pencil is $\frac{51}{10}$ of a meter long. What is the length, in meters, of Peter's pencil written as a decimal?

A. 5.1

B. 0.051

C. 0.51

D. 51.10

17) Mia has a group of shapes. Each shape in her group has at least one set of parallel sides. Each shape also has at least one set of perpendicular sides. Which group could be Mia's group of shapes?

A.

B.

C.

D.

18) There are 58 students from Riddle Elementary school at the library on Monday. The other 17 students in the school are practicing in the classroom. Which number sentence shows the total number of students in Riddle Elementary school?

A. 58 + 17

B. 58 − 17

C. 58 × 17

D. 58 ÷ 17

19) A stack of 4 pennies has a height of 1 centimeter. Elise has a stack of pennies with a height of 16 centimeters. Which equation can be used to find the number of pennies, n, in Elise's stack of pennies?

A. n = 16 + 4

B. n = 16 − 4

C. n = 4 × 16

D. n = 16 ÷ 34

20) Use the models below to answer the question.

Which statement about the models is true?

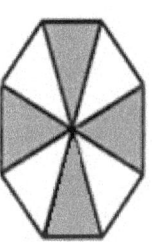

A. Each shows the same fraction because they are the same size.

B. Each shows a different fraction because they are different shapes.

C. Each shows the same fraction because they both have 4 sections shaded.

D. Each shows a different fraction because they both have 4 shaded sections but a different number of total sections.

21) What is the value of A in the equation 24 ÷ A = 4?

A. 2

B. 6

C. 7

D. 8

STAAR Subject Test Mathematics Grade 4

22) Sophia flew 7,143 miles from Los Angeles to New York City. What is the number of miles Sophia flew rounded to the nearest thousand?

A. 7,000

B. 7,100

C. 7,140

D. 8,000

23) Write $\frac{75}{100}$ as a decimal number.

A. 7.05

B. 0.75

C. 7.5

D. 0.075

24) Erik made 12 pints of juice. He drinks 2 cups of juice each day. How many days will Erik take to drink all of the juice he made?

A. 14 days

B. 6 days

C. 10 days

D. 4 days

STAAR Subject Test Mathematics Grade 4

25) Jason has prepared $\frac{8}{1,000}$ of his assignment. Which decimal represent the part of the assignment Jason has prepared?

   A. 8.110

   B. 8.001

   C. 0.008

   D. 0.0008

26) Emma described a number using these clues.

   ✓ 3 digits of the number are 9, 3, and 2

   ✓ The value of the digit 3 is (3 × 10)

   ✓ The value of the digit 2 is (2 × 1,000)

   ✓ The value of the digit 9 is (9 × 10,000)

   Which number could fit Emma's description?

   A. 92,563.10

   B. 29,563.70

   C. 92,536.17

   D. 92,635.71

27) There are 12 boxes and each box contain 24 pencils. How many pencils are in the boxed in total?

   A. 188

   B. 208

   C. 288

   D. Not here

STAAR Subject Test Mathematics Grade 4

28) Emily and Ava were working on a group project last week. They completed $\frac{7}{17}$ of their project on Tuesday and the rest on Wednesday. Ava completed $\frac{6}{17}$ of their project on Tuesday. What fraction of the group project did Emily completed on Tuesday?

   A. $\frac{2}{17}$

   B. $\frac{13}{17}$

   C. $\frac{1}{17}$

   D. $\frac{5}{17}$

29) Moe packs 50 boxes with flashcards. Each box holds 80 flashcards. How many flashcards Moe can pack into these boxes?

   A. 4,000

   B. 400

   C. 700

   D. 4,400

30) Jason's favorite sports team has won 0.13 of its games this season. How can Jason express this decimal as a fraction?

   A. $\frac{1}{13}$

   B. $\frac{13}{10}$

   C. $\frac{13}{100}$

   D. $\frac{0.13}{100}$

"This is the end of Practice Test 1"

State of Texas Assessments of Academic Readiness

STAAR Practice Test 2

Mathematics

GRADE 4

Released Month Year

STAAR Subject Test Mathematics Grade 4

1) Which statement about the number 495,271.38 is true?

   A. The digit 9 has a value of (9 × 1,000)

   B. The digit 5 has a value of (5 × 100)

   C. The digit 7 has a value of (7 × 10)

   D. The digit 4 has a value of (4 × 10,000)

2) What is the perimeter of this rectangle?

   A. 16 cm

   B. 28 cm

   C. 14 cm

   D. 24 cm

3) What is the eighth number in the following pattern?

   1,450; 1,700; 1,950; 2,200; ____, ____, ____, ____

   A. 3,400

   B. 3,150

   C. 3,350

   D. 3,200

4) Jamie has 4 quarters, 8 dime, and 25 pennies. How much money does Jamie have?

   A. 195 pennies

   B. 205 pennies

   C. 205 pennies

   D. 190 pennies

5) Jeb paid $96 for a magazine subscription. If he is paying $8 for each issue of the magazine, how many issues of the magazine will he receive?

   A. 12

   B. 14

   C. 24

   D. 10

6) What is the perimeter of the triangle?

   A. 46 inches

   B. 68 inches

   C. 84 inches

   D. 92 inches

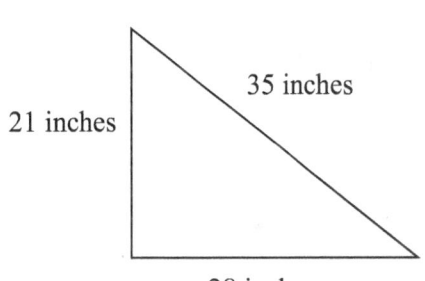

7) Emma draws a shape on her paper. The shape has four sides. It has both diagonals have equal measure and are perpendicular. What shape does Emma draw?

   A. parallelogram

   B. rectangle

   C. square

   D. trapezoid

STAAR Subject Test Mathematics Grade 4

8) Order the fractions from least to greatest. $\frac{5}{8}, \frac{1}{2}, \frac{5}{6}, \frac{1}{24}$

   A. $\frac{5}{8}, \frac{5}{6}, \frac{1}{2}, \frac{1}{24}$

   B. $\frac{5}{6}, \frac{5}{8}, \frac{1}{24}, \frac{1}{2}$

   C. $\frac{1}{24}, \frac{1}{2}, \frac{5}{8}, \frac{5}{6}$

   D. $\frac{1}{2}, \frac{1}{24}, \frac{5}{6}, \frac{5}{8}$

9) There were 52 students in the first row and 9 students in the second row. How many students were in the first two rows?

   A. 61

   B. 57

   C. 43

   D. 55

10) Jack has $31.00. He earns $14.00 more. How much money does Jack have in all?

   A. $54.00

   B. $44.00

   C. $46.00

   D. $45.00

STAAR Subject Test Mathematics Grade 4

11) Joe put 15 red cards and 6 black cards in each bag. What is the total number of cards Joe put in 4 bags?

Write your answer in the box below.

12) Rounded to the nearest 100,000, the population of Louisiana was 54,700,000 in 2010. Which number could be the actual population of Louisiana in 2010?

A. 54,789,221

B. 54,752,220

C. 54,723,330

D. 54,762,990

13) A football teams play 48 games each year. How many games will the team play in 5 years?

A. 260

B. 180

C. 240

D. 1,100

14) The figure below shows a diagram of a living room.

The perimeter of the living room is 26 feet (ft). What is the width(w) of the living room?

A. 12 ft

B. 7 ft

C. 5 ft

D. 11 ft

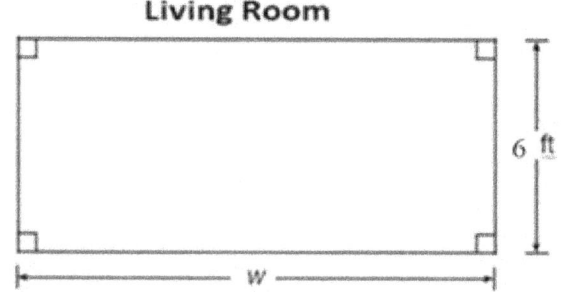

15) A building is 63 feet high. What is the height of the building in yards?

A. 22 yards

B. 21 yards

C. 23 yards

D. 66 yards

16) The sum of A and B equals 35. If A = 18, which equation can be used to find the value of B?

A. B – 18 = 35

B. B + 18 = 35

C. A + 18 = 35

D. A – 18 = 35

STAAR Subject Test Mathematics Grade 4

17) Which number is represented by A? $4 \times A = 172$

    A. 43

    B. 37

    C. 47

    D. 31

18) Which fraction has the least value?

    A. $\frac{1}{32}$

    B. $\frac{1}{2}$

    C. $\frac{3}{4}$

    D. $\frac{5}{8}$

19) Which triangle has one obtuse angle?

A.

C.

B.

D.

STAAR Subject Test Mathematics Grade 4

20) Lisa has 64 pastilles. She wants to put them in boxes of 16 pastilles. How many boxes does she need?

   A. 8

   B. 5

   C. 6

   D. 4

21) What is the volume of the cube?

   Write your answer in the box below.

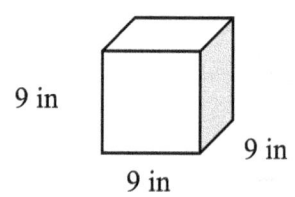

22) To what number is the arrow pointing?

   A. 16

   B. 14

   C. 12

   D. 22

23) What mixed number is shown by the shaded rectangles?

   A. $3\frac{3}{4}$

   B. $3\frac{1}{4}$

   C. $2\frac{1}{4}$

   D. $2\frac{3}{4}$

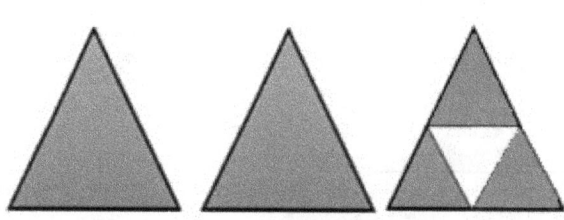

STAAR Subject Test Mathematics Grade 4

24) A straight-line measure 180°. A straight line and a triangle are touching as shown in the figure below.

What is the value of A in the figure?

A. 60

B. 20

C. 90

D. 110

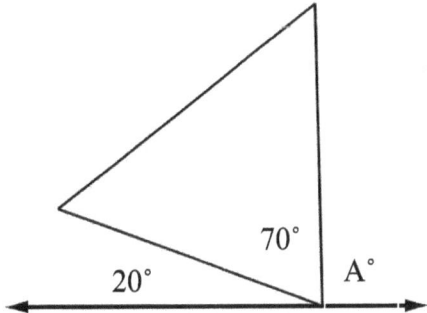

25) The number 64.05 can be expressed as _____

A. $(6 \times 10) + (4 \times 1) + (5 \times 0.01)$

B. $(6 \times 10) + (4 \times 1) + (5 \times 0.1)$

C. $(6 \times 1) + (4 \times 1) + (0 \times 1) + (5 \times 1)$

D. $(6 \times 10) + (4 \times 1) + (0 \times 10) + (5 \times 100)$

26) What is the perimeter of this shape?

A. 72

B. 54

C. 30

D. 33

STAAR Subject Test Mathematics Grade 4

27) The temperature on Sunday at 12:00 PM was 92°F. Low temperature on the same day was 53°F cooler. Which temperature is closest to the low temperature on that day?

   A. 58°F

   B. 39°F

   C. 36°F

   D. 48°

28) Which shape shows a line of symmetry?

   A.

   C.

   B.

   D.

29) There are 365 days in a year, and 24 hours in a day. How many hours are in one-sixth years?

   A. 1,340

   B. 2,140

   C. 1,920

   D. 1,460

STAAR Subject Test Mathematics Grade 4

30) On Sunday Leon was a referee at 4 soccer games. He arrived at the soccer field 20 minutes before the first game. Each game lasted for 20 minutes. There were 10 minutes between each game. Leon left 15 minutes after the last game. How long, in minutes, was Leon at the soccer field?

A. 215 minutes

B. 145 minutes

C. 155 minutes

D. 95 minutes

"This is the end of Practice Test 2."

STAAR Subject Test Mathematics Grade 4

STAAR Subject Test Mathematics Grade 4

Chapter 13 : Answers and Explanations

STAAR Practice Tests Answer Key

✷ Now, it's time to review your results to see where you went wrong and what areas you need to improve!

STAAR- Mathematics

Practice Test - 1

1	D	11	D	21	B
2	B	12	B	22	A
3	C	13	B	23	B
4	680	14	B	24	B
5	A	15	A	25	C
6	C	16	A	26	C
7	C	17	B	27	C
8	C	18	A	28	C
9	B	19	C	29	A
10	C	20	D	30	C

Practice Test - 2

1	C	11	84	21	$729\ in^3$
2	D	12	C	22	B
3	D	13	C	23	D
4	C	14	B	24	C
5	A	15	B	25	A
6	C	16	B	26	D
7	C	17	A	27	B
8	C	18	A	28	A
9	A	19	A	29	D
10	D	20	D	30	B

STAAR Subject Test Mathematics Grade 4

Practice Test 1
STAAR- Mathematics
Answers and Explanations

1) Answer: D.

$50 \times 32 = 1,600$.

2) Answer: B.

Tam wants to divide his 350 cards into boxes of 70 cards. Therefore, he needs $350 \div 70 = 5$ boxes.

3) Answer: C.

The clock shows 5:15 in the noon. 2 hours ago, it was 3:15 AM. 45 minutes before that was 2:30 AM.

4) Answer: 680.

We round the number up to the nearest ten if the last digit in the number is 5,6,7,8, or 9
We round the number down to the nearest ten if the last digit in the number is 1,2,3, or 4
If the last digit is 0, then we do not have to do any rounding, because it is already to the ten. Therefore, rounded number of 679 to the nearest ten is 680.

5) Answer: A.

Use area of rectangle formula. Area = length × width

The length of the rectangle is about 4 inches and its width are about 3 inches.

Therefore, the area of the rectangle is $4 \times 3 = 12$ inches.

6) Answer: C.

7 days = 1 week, 31 days = $(31 \div 7) = 4.29 \Rightarrow 4$ weeks

7) Answer: C.

The football team should buy 15 uniforms and each uniform cost $19. Therefore, they should pay $(15 \times \$ 19)$ $285.

Choice C is correct answer:

$(19 \times 10) + (19 \times 5) = 190 + 95 = \$285$

STAAR Subject Test Mathematics Grade 4

8) Answer: C.

20 + Z = 55. Therefore, Z = 55 – 20 = 35

Let's review the options provided.

A. 55 – 20 = Z	Yes! This is true. 55 – 20 = 35
B. 55 – Z = 20	Yes! This is true. 55 – 35 = 20
C. Z – 20 = 55	No! This is not true. 35 - 20 = 15
D. Z + 20 = 55	Yes! This is true. 35 + 20 = 55

Option C is the only option that is NOT true.

9) Answer: B.

Three – digit even numbers that have a 5 in the hundreds place and a 3 in the tens place are 530, 532, 534, 536, 538.

Option B, 532, is the answer.

10) Answer: C.

The number of votes for Football was 9, for basketball was 4, for soccer was 10, and for volleyball was 2. Only table C shows all these numbers correctly.

11) Answer: D.

There are 100 equal parts. 66 parts are shaded. It is equal to $\frac{66}{100}$ or 0.66.

12) Answer: B.

Choice B shows the numbers in order from least to greatest.

21,851 < 28,081 < 30,032 < 37,009

13) Answer: B.

6.0 – 4.75 = 1.25

14) Answer: B.

Add adult tickets and children's ticket that have been sold.

$\frac{52}{100} + \frac{1}{10} = \frac{52}{100} + \frac{10}{100} = \frac{52+10}{100} = \frac{62}{100} = \frac{31}{50}$

15) Answer: A.

This angle is less than 180° and more than 90°. Only choice A.

WWW.MathNotion.Com

16) Answer: A.

$\frac{51}{10}$ is equal to 5.1.

17) Answer: B.

Only on option B, all shapes have at least one set of parallel side and one set of perpendicular side.

18) Answer: A.

Add the number of students at the library and number of students in classroom.
$58 + 17 = 75$

19) Answer: C.

For the height of 1 centimeter, we have 4 pennies, therefore, for the height of 16 centimeters, we have 4×16 pennies.

20) Answer: D.

The first model from left is divided into 6 equal parts. 4 out of 6 parts are shaded. The fraction for this model is $\frac{4}{6} = \frac{2}{3}$. The second model is divided into 8 equal parts. 4 out of 8 parts are shaded. Therefore, the fraction of the shaded parts for this model is $\frac{4}{8} = \frac{1}{4}$. These two models represent different fractions.

21) Answer: B.

$24 \div A = 4$, therefore, $A = 24 \div 4 \rightarrow A = 6$

22) Answer: A.

When rounding to the nearest thousand, you will need to look at the last three digits. If the last three digits is 499 or less round down the number ending with three zeros. On the other hand, if the last three digits is 500 or more, round up the next ending with three zeros. Since, in the number 7,143, the number 143 is smaller than 499, then round the number to 7,000.

23) Answer: B.

$\frac{75}{100}$ is equal to 0.75.

24) Answer: B.

$12 \div 2 = 6$.

STAAR Subject Test Mathematics Grade 4

25) Answer: C.

$\frac{8}{1,000}$ is equal to 0.008.

26) Answer: C.

Only option C fits Emma's description.

27) Answer: C.

$12 \times 24 = 288$.

28) Answer: C.

$\frac{7}{17} - \frac{6}{17} = \frac{1}{17}$.

29) Answer: A.

If one box has 80 flashcards, therefore, 50 boxes have the capacity of (50 × 80) 4,000 flashcards.

30) Answer: C.

0.13 is equal to $\frac{13}{100}$.

Practice Test 2
STAAR- Mathematics
Answers and Explanations

1) Answer: C.

A. The digit 9 has a value of 9 × 10,000, not 9 × 1,000.

B. The digit 5 has a value of 5 × 1,000, not 5 × 100.

C. The digit 7 has a value of 7 × 10. This is true!

D. The digit 4 has a value of 4 × 100,000, not 4 × 10,000.

2) Answer: D.

use perimeter of rectangle formula. P = 2 (length + width)

P = 2 × (4 + 8) = 2 × 12 = 24 cm

3) Answer: D.

The difference of each two successive numbers is 250.

Add four 250 to last number (2,200):

2,200 + 250 + 250 + 250 + 250 = 3,200

4) Answer: C.

4 quarters = 4 × 25 pennies = 100 pennies

8 dimes = 8 × 10 pennies = 80 pennies

In total Nicole has 205 pennies

5) Answer: A.

1 issue of the magazine = $8.

$96 ÷ $8 = 12 issues of the magazine.

6) Answer: C.

To find the perimeter of a triangle, add all three sides of the triangle.

P = 21 + 28 + 35 = 84 inches

7) Answer: C.

In any square, both diagonals have equal measure and are perpendicular.

STAAR Subject Test Mathematics Grade 4

8) Answer: C.

To compare fractions, we can write fractions with the same denominator. Then, we can compare the numerators of each fraction and put them in correct order from least to greatest or greatest to least.

Common denominator of 8, 2, 6 and 24 is 24. Rewrite the fractions:

$\frac{5}{8} = \frac{15}{24}$ $\frac{1}{2} = \frac{12}{24}$ $\frac{5}{6} = \frac{20}{24}$ $\frac{1}{24} = \frac{1}{24}$

9) Answer: A.

$52 + 9 = 61$.

10) Answer: D.

$31 + $14 = $45.

11) Answer: 84.

1 bag = 15 red cards + 6 black cards (15 + 6 = 21 cards)

4 bags = 4 × 21 = 84 cards.

12) Answer: C.

To round numbers to the nearest ten thousand, make the numbers whose last five digits are 00001 through 49999 into the next lower number that ends in 00000. For example, 724,424 rounded to the nearest one hundred thousand would be 700,000. Choice C is correct because last five digits of 54,723,330 is less than 49,999.

13) Answer: C.

1 year = 48 games. 5 years = 5 × 48 = 240 games

14) Answer: B.

Use perimeter of rectangle formula. P = 2 × (Length + Width)

26 = 2 × (6 + W) ⇒ 26 = 12 + 2 × W ⇒ 2 × W = 26 − 12 = 14 ⇒ W = 7 feet.

15) Answer: B.

3 feet = 1 yard;   63 feet = 21 yards.

16) Answer: B.

The sum of A and B equals 35: A + B = 35, if A = 18, then B = 35 − 18 = 17.   18 + 17 = 35 or B + 18 = 35

STAAR Subject Test Mathematics Grade 4

17) Answer: A.

A = 172 ÷ 4 = 43

18) Answer: A.

To compare fractions, we can write fractions with the same denominator. Then, we can compare the numerators of each fraction and put them in correct order from least to greatest or greatest to least.

Common denominator of 2, 4, 8, and 32 is 32. Rewrite the fractions:

A. $\frac{1}{32}$ B. $\frac{1}{2} = \frac{16}{32}$ C. $\frac{3}{4} = \frac{24}{32}$ B. $\frac{5}{8} = \frac{20}{32}$

19) Answer: A.

An obtuse triangle is one with one obtuse angle (greater than 90°) and two acute angles. Since a triangle's angles must add up to 180°, no triangle can have more than one obtuse angle.

Only shape A has one obtuse angle.

20) Answer: D.

16 pastilles = 1 box

64 pastilles = (64 ÷ 16) = 4 boxes

21) Answer: 729.

To find the volume of cube, multiply one side of the cube by itself 3 times:

Volume of a cube = (side) × (side) × (side) = 9 × 9 × 9 = 729 in^3.

22) Answer: B.

Arrow is pointing to a number in the middle of two numbers 10 and 18. Therefore, the answer is 14.

23) Answer: D.

This shape shows 2 complete shaded triangles and a three of a four part of triangle.

The mixed number for this shape is $2\frac{3}{4}$.

24) Answer: C.

Three angles in a triangle add up to 180°.

A° + 70° + 20° = 180° ⇒ A° = 180° − 90° = 90°

STAAR Subject Test Mathematics Grade 4

25) Answer: A.

The number 64.05 can be expressed as:

$(6 \times 10) + (4 \times 1) + (5 \times 0.01) = 60 + 4 + 0.05 = 64.05$

26) Answer: D.

To find the perimeter of the shape, add all four sides.

$P = 6 + 12 + 6 + 9 = 33$

27) Answer: B.

Low temperature is 53°f cooler than the temperature at 12:00 PM which is 92°. Low temperature is 39°f (92°f – 53°f) that is choice B.

28) Answer: A.

You can find if a shape has a Line of Symmetry by folding it. When the folded part sits perfectly on top (all edges matching), then the fold line is a Line of Symmetry. Shape A shows a line of symmetry.

29) Answer: D.

1 year = 365 days, 1 day = 24 hours

$\frac{1}{6}$ years = $\frac{1}{6} \times 365 \times 24 = 1,460$ hours

30) Answer: B.

Each game 20 minutes, therefore 4 games took $4 \times 20 = 80$ minutes.

10 minutes between each game. There are 30 minutes in total between 4 games. (between game 1 and 2, 10 minutes and between game 2 and 3, 10 minutes and between game 3 and 4, 10 minutes)

Leon arrives 20 minutes before first game and left 15 minutes after the last game. In total, he was 80 + 30 + 20 + 15 = 145 minutes at the soccer field.

"End"

www.ingramcontent.com/pod-product-compliance
Lightning Source LLC
Chambersburg PA
CBHW080442110426
42743CB00016B/3249